Star

of

Greece

For Profit and Glory

Final Edition

Paul W Simpson

Clippership Press

Star of Greece

Also by the same author.

Around Cape Horn Once More
Windjammer – Tales of the clipper ship Loch Sloy
Neptune's Car
'Ole Virginia Bound
The Last Captain

**All books available at
these online retail outlets.**

Star of Greece

Copyright.

Clippership Press.
Adelaide, South Australia
paulsimo2010@yahoo.com.au

© Paul W Simpson 2020

National Library of Australia Cataloguing-in-Publication entry

Author: Simpson, Paul W.,(1969 -) author.

Title: Star of Greece – For Profit & Glory / Paul W Simpson.

ISBN: 978-1-71663-257-0
Final Print Edition.

Notes: Includes bibliographical references and index.

Subjects: Star of Greece (Ship)
Shipwrecks--South Australia--History.
Sailing ships--Ireland--History
Sailing ships--India--History.
Sailing ships--Australia--History.

Cover Illustration:

Star of Greece

Dedication.

**To Fon. My guiding star upon a smooth and
mirrored sea.**

*"The sea - this truth must be
confessed –
has no generosity.
No display of manly qualities –
courage, hardihood, endurance,
faithfulness –
has ever been known to touch
its irresponsible consciousness of
power."*

Joseph Conrad

Star of Greece

Foreword

The clipper ship 'Star of Greece' was launched in 1864 from the yards of Harland & Wolff in Belfast, Northern Ireland. She was one of a fleet of iron vessels built for J.P. Corry, jute and timber importers for the Calcutta jute trade. This book is the story of the ship and the trade in which she sailed during the latter half of the nineteenth century. The vessels career was first covered by Basil Lubbock in his book, 'The Last of the Windjammers' Volume One (1948). The 'Star of Greece' plied the ways between Britain, India and Australia for twenty-four years before coming to grief at Port Willunga, South Australia, in the early hours of July 13th 1888. The demise of the ship and eighteen of her crew has been dramatically covered by the authors Rae Sexton and Geoffrey H. Manning. And the members of the Willunga Branch of the National Trust of South Australia in their wonderful books 'Before the Wind' (1982), 'The Tragic Shore' (1988) and 'Star of Greece and other shipwrecks of the tragic shore' (2018).

In 2018 upon the 130th anniversary of the loss of the Star of Greece, a large group gathered at Aldinga cemetery to pay their respects to the dead. After a solemn ceremony conducted by a local minister, a plaque was unveiled with the names of all who had died. The unstinting efforts of the members of Willunga Branch of the National Trust of South Australia had raised the funds for the new memorial. Now when you visit the Star of Greece obelisk located within the Aldinga Uniting Church Cemetery you will find a modest addition alongside. Upon this bronze, plate is the names of all the men and boys who are known

Star of Greece

to have perished aboard the Star of Greece when she came ashore at Port Willunga on July 13th 1888.

The original purpose of this book was to restore to history the stories of the Star of Greece and to bring to light the lives of the 28 crew. Now thanks in large part to the efforts of the Star of Greece Working Group (Willunga Branch of the National Trust), the names of her crew have been restored to history. Now perhaps the ghosts of the lost can rest a little easier knowing that they are no longer amongst the lost and forgotten. Special thanks must go to Mark Staniforth and Paddy O'Toole for bringing the idea of a memorial plaque to life.

There is now a 'Star of Greece' heritage trail that covers a vast swath of Adelaide and the Southern Vales'; starting at the South Australian Maritime Museum to the Semaphore Sea Captains Memorial, on to the National Trust Museum at Willunga, the Aldinga Public Library, Port Willunga Beach where the Star of Greece was wrecked and finishing at the Aldinga Uniting Church Cemetery, Old Coach Road, Aldinga. Take this book with you as you follow the trail. Relive those final dramatic days of the final voyage of the ill-starred clipper – The Star of Greece.

Paul W Simpson 2020.

Star of Greece

Star of Greece

Star of Greece

Rise of the Red Heart Line

The chief partners of Harland & Wolff, Gustav Wilhelm Wolff, Walter Henry Wilson, William James Pirrie and Edward James Harland.

The sea is a fickle mistress, unforgiving and indifferent to the fates of those who choose to sail upon her. Yet she is also a provider of great bounty for those who dare her deep waters. Much has been written already about the famous Irish company J.P. Corry & Co'. For decades importers of timbers from North America and the Baltic there came a time when the company decided to 'cut out the middle man' and go into the timber shipping business for themselves. The freight line was established in 1826 by timber merchant Robert Corry with the purchase of the 325-ton wooden ship 'Chieftain' for the import of timber. Corry's soon purchased more wooden hulls for use in the importation of Canadian timber, and jute from India. His other wooden-hulled ships included the Queen of the West, Alabama, Charger, Persian, St Helena and Summerhill. The Queen of the West was lost off South Africa in 1850 on the run from Bombay to London

Star of Greece

going down with all 32 of her crew. In 1851, Robert's son, James joined the company and they became Corry & Co'. Often their wooden ships were manufactured in Canada then on their maiden runs, loaded down with timber for Belfast and Liverpool.

In 1859 the Charger was placed on the run to Calcutta to bring home a load of Jute, linseed, and sacking. By the late 1850s, there was a move away from wood to the more resilient iron hulls. There was also the problem of jute and linseed being highly flammable, a very real danger for wooden ships at sea. With this in mind, Corry's placed an order with Belfast shipbuilders Harland & Wolff for an iron-hulled clipper specifically designed to carry general cargo or coal to India and bales of jute and linseed back to London, Dundee and Liverpool. The order for the new clipper was placed at the same time as the 1000-ton Charger set sail for India in 1859. Despite the dangers there existed within men's souls a constant yearning for profit, adventure and glory, a rare brew only attainable by the few who chose a life at sea.

The result was a beautiful example of an iron clipper, the first of many to be built for Corry & Co' by Harland & Wolff. At 953 ton, the newly finished ship came in at 200 feet long, with a beam of 32 feet and a 21-foot draft. Fire at sea was the sailor's greatest fear, therefore, the little clipper came with three bulkheads to keep the burning portion of a cargo isolated from the rest of the ship. This was to be a regular feature in all Corry's iron sailing ships. Whilst none were lost to fire in over 20 years, the same design feature would prove fatal for many of those who sailed in a Corry ship. The new clipper christened 'Jane Porter' was named after William Corry's wife. The ship was taken from the stocks of Harland & Wolff's Belfast yard by Captain J. McDowell who sailed her across to Glasgow to load with general cargo for Calcutta. From Calcutta, the clipper, her hold and 'tween deck spaces crammed with bales of jute sailed for London. The voyage proved profitable enough for Corry & Co' to decide to invest heavily in the growing jute and linseed trades.

This investment resulted in the ordering of 12 similarly designed iron clippers from Harland & Wolff. The first of these to be finished was the beautiful 'Star of Erin', at 949 tons and 200 feet in length, with

Star of Greece

Wilhelm Wolff designer of each of the Star Liners except the Star of Austria.

Sir Edward J Harland.
Founder of shipbuilding firm Harland & Wolff.

Sir James Porter Corry M.P.
Artist - William Ewart Lockhart.
Ulster Museum.

Star of Greece

three transverse bulkheads she was a virtual sister-ship to the Jane Porter. Ordered in 1861, she was launched in early 1862 and sailed from Belfast to Calcutta via London. She was followed in short order by the slightly larger (200 foot long, 998 tons) 'Star of Denmark'. Launched in 1863 she voyaged from Liverpool to Calcutta and came with the obligatory three internal bulkheads. These three vessels were followed by the launch in 1864 of the 999-ton sisters 'Star of Albion' and 'Star of Scotia'. Corry's interests in the trade with India continued to grow as demand for Jute, gunny sacking and linseed increased in Britain. To meet this demand Corry's ordered a pair of identical 1200 Jute clippers from Harland & Wolff in 1867.

The twin full-rigged ships were to be the finest ever launched from Harland & Wolff's yard up to that point. The first of this new breed of vessels to come off the slipway at Queens Island on June 23rd 1868 was named the 'Star of Persia', in line with the naming habit of the rest of the Star Line. With a keel length of 225 feet, a beam of 35 feet and a draft of 22 feet, the brand new clipper came in at a little over 1200 tons. Her launch was witnessed by a great number of spectators and was accompanied by much fanfare. A festive air surrounded the ship as commoners, workers and dignitaries alike enjoyed the occasion as a brass band entertained the throng. The ship was duly christened with the usual pomp and ceremony by the wife of the owner, Mrs John Corry.

Harland & Wolff then entertained the notables at a lengthy luncheon at which many congratulatory toasts were made. Mr Edward Harland said; *"Success to the Star of Persia. She is the sixth vessel with the construction of J.P. Corry & Co'....it is the wish of everyone present – that the success which has attended her predecessors will not be wanting to the Star of Persia...and while some shipowners have hesitated about investing more capital in ships, this enterprising firm have looked forward to better days, and ordered not only the one ship, but two ships, for the vessel you have seen launched today is only the forerunner of a second, which we hope you will see launched in a month or six weeks."*
Belfast News-Letter 24 June 1868

Star of Greece

Jane Porter. Circa 1880's Unknown Artist.
National Museums Northern Ireland.

Star of Greece

Star of Persia. Circa 1900 Unknown Artist.
National Museums Northern Ireland.

Star of Greece

Star of Germany
Brodie Collection, State Library of Victoria.

Star of Greece

Star of Russia
Brodie Collection, State Library of Victoria.

Star of Greece

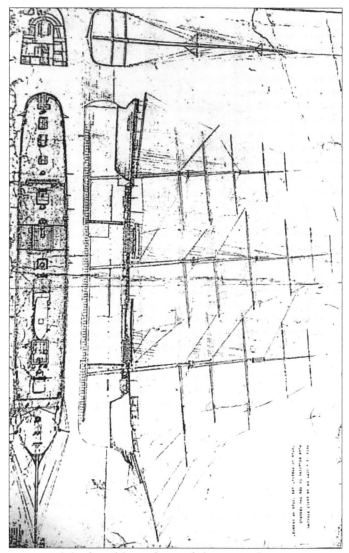

Original Builders Plans of the Star of Greece
Harland & Wolff Shipyards, circa 1868.

Star of Greece

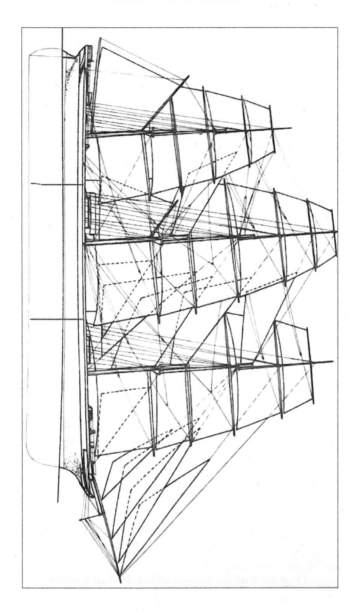

Star of Greece

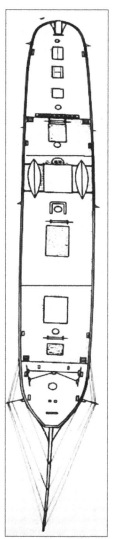

Sail and deck plans of Star of Greece, taken from Harland & Wolff builder's plans, circa 1868.
Harland & Wolff archives, Belfast.

Star of Greece

With the festivities over for the afternoon, the newly launched iron clipper was hauled from Princes Dock over to Abercorn Basin so that the masts, jib-boom, gaffs and yards could be fitted and the rigging completed.

The Belfast shipbuilders were as good as their word when in mid-September Gustav Wolff and Edward Harland had invitations sent out to J.P. Corry and their associates inviting them to the launch of their latest clipper and obligatory post-launch luncheon. The usual party of dignitaries and a large crowd of spectators gathered at the Queens Island slipway at high water on Saturday, September 19th 1868, to witness the launch of Corry's latest acquisition. In almost every aspect the shiny new clipper with her matt black strakes, gleaming white bulwarks, and red boot-top was the spitting image of her sister the 'Star of Persia' which had recently sailed for Calcutta. Just as with the previous launch, the new ship was christened the 'Star of Greece' by Robert Corry's daughter. It had originally been proposed to call the new vessel 'Queen of Greece', but this was not in keeping with the rest of Corry's fleet. The launch luncheon was filled with laughter and much frivolity as high hopes were expressed by all parties for the future success of both the 'Star of Greece', J.P. Corry and Harland & Wolff. With the preliminaries dealt with the ship was towed from Prince's Dock to Abercorn for rigging and final fit-out.

There were six more clippers constructed by Harland & Wolff for J.P. Corry's 'Calcutta Star Line'. These were the 1337 ton, 232 foot long 'Star of Germany' launched in 1872, the 1372 ton, 262 foot, 'Star of Bengal, coming out of Queens Island alongside her sister the 'Star of Russia' in 1874. The next two clippers off the stocks were the 1644 ton, 272 foot long clippers, 'Star of Italy' and 'Star of France'. The last of Corry's clippers was built by a different firm as Harland & Wolff turned their attention to the building of steam-driven leviathans. This last vessel, the 'Star of Austria' came in at 264 feet long and weighed 1781 tons. She was built for the amount of cargo she could carry rather than the speed of her passages, and her lines reflected this thinking. The ship, built by Workman, Clark & Co', of Belfast in 1886, was the last sailing ship built for Corry's before they sold off all their remaining deep-water sailers and moved wholly into steam.

Star of Greece

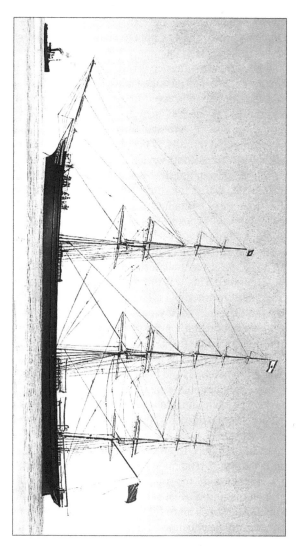

Photograph of a watercolour painting of the Star of Greece.
Artist J. Kerr. State Library of Victoria.

Star of Greece

The final rigging, fit-out and preparation for the sailing of the 'Star of Greece' were superintended by her recently appointed skipper, William James Munce Shaw. Captain Shaw was a doughty, hard-driving Irishman from Ballyshannon. Born in 1837, the 31-year-old Captain had been a ship's master for the last seven years and came with the reputation as a fast passage maker and sail carrier who drove his ship and the sailors who served under him, very hard. The 'Star of Greece' like her sister was heavily rigged with iron masts and lower yards. She initially crossed three skysail yards over single topgallants and double topsails. This plan was later cut down due to her dismasting on her second voyage and she henceforth carried a skysail just upon her mainmast. Most of the crew were housed in a topgallant fo'c'sle, whilst the idlers were accommodated in a mid-deck house that also contained the carpenter's and sailmaker's workshops, and bosun's locker.

The apprentices had the worst of it housed as they were in the half-deck. This house positioned abaft the mainmast also held the galley and a donkey engine. This was the wettest place on the ship to sleep and was frequently washed out. The high poop with its whale-back end housed the officers and any passengers in relative luxury. The Captains' single berth cabin on the port side came with a wide bunk atop inbuilt drawers for clothing. It also contained a desk, chart table, cupboards and a wash-stand (with mirror and running water). Off to one side was a private toilet and bathroom. The cabin was lit by twin portholes and gimballed kerosene or oil lanterns. The first officer's cabin was the same but situated upon the starboard side of the ship. All other cabins were twin-berth cells with just enough room for a sea chest and a washstand. Each cabin also had a toilet and there was a bath right aft in the poop. In all, there were five cabins for officers, guests and the steward. The main eating and entertainment area was the saloon. Lined with red maple and mahogany it was lit by gimballed lanterns and a great skylight fitted with windows on the side. These allowed fresh air to fill the cabin when weather permitted. The clipper was not designed to be a passenger liner and thus carried few other luxuries or entertainments. The final fit-out and rigging of the ship were completed by the first week of October, and soon she was made ready for her tow across to Liverpool to take on coal for the jute mills and railways of Calcutta.

Star of Greece

Racing & Reputations

Captain Shaw was at last satisfied that his ship was ready for sea. The tug towed her from Abercorn Basin and out into the River Lagan. Officials from J.P. Corry and Harland & Wolff were on hand to witness the clippers departure. With a final hoot of the tugs steam whistle and cheers from the shore, the 'Star of Greece' disappeared down river and out to sea. A small crew of ships runners and riggers had stayed aboard to fine-tune the rigging before she put out to sea. The trip across the channel and up the Mersey was free from trouble and the 'Star of Greece' was hauled into the docks to begin loading. In less than 10 days her cargo had been successfully stowed and the final preparations were being made even as the last of the crew were being rousted from their boarding houses and dockside taverns. The ship was entered out on October 21st and towed out to sea the next morning. From the Holy Head, she left her tug and set sail through St George's Channel pushed along by a freshening north-westerly breeze. Sailing conditions improved as the clipper cleared the Bay of Biscay with the Cape Verde Island being passed on November 10th after just 20 days at sea. Christmas was spent thrashing around the Cape of Good Hope into northeast to southeast winds and confused seas.

The real risk was running into the southern Indian Ocean at the height of the cyclone season. Captain Shaw's fears were not realised as the Star of Greece humming along on the southeast trades ran straight into a series of baffling calms and headwinds. Fickle winds and currents of the Bay of Bengal in January made for a fine but slow voyage to Calcutta. The muddy waters of the Hooghly River began to discolour the Bay indicating that they were approaching the shifting sandbars, shoals and channels of the Hooghly mouth. Bill Shaw checked his charts and kept the lead going as lookouts maintained a weather eye for a pilot schooner. Saugor light was sighted on the evening of Friday, January 29th as Captain Shaw hove his ship to and dropped anchor in the outer roadstead 100 days from Liverpool. With the pilot aboard on the 30th a steam tug was engaged and the 'Star of Greece'

Star of Greece

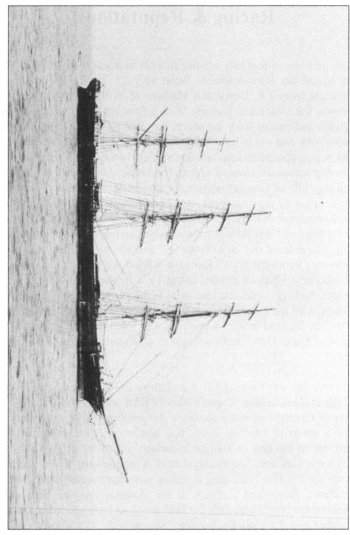

The Star of Greece at anchor.
The Advertiser Newspaper, circa 1938.
State Library of Victoria.

Star of Greece

was towed to Diamond Harbour to await the next flood tide. Having successfully navigated the congested waterway that was the lower Hooghly the crew of the clipper tied up to the mooring buoy on February 2nd, in the heart of Garden Reach to await a berth alongside Calcutta's busy docks.

The demand for jute was growing when J.P Corry decided to diversify their import base. The jute would come in from Calcutta in bales and be sent by rail from London to Dundee and other mills after having passed through the London sales. With the Star of Greece's coal successfully discharged, enough ballast was taken aboard to prevent her capsizing as she was towed upriver to the jute market at Hautkolo. Here the great jute presses and screw houses were also situated and the clipper was loaded from cargo lighters. Stevedores worked feverishly in the dark and stiflingly hot confines on the ships holds to twist in as many bales as could be safely stowed.

With such industrial efficiency, it was but a month from the time of the Star of Greece's arrival that she was fully loaded down to her marks and again ready to put to sea. She was cleared out from Hautkolo on February 28th and dragged back down to Garden Reach to await a pilot and a steamer to become available. With a tow arranged after some spirited haggling by Captain Shaw, the Star of Greece was towed down on the flood tide to anchor once again off Diamond Head. The next day with enough water over the bar the clipper slipped through the main channel and dropped her pilot off of Saugor on March 2nd 1869. Shaw's crew were in high spirits for their rival, the Star of Scotia had departed Saugor on the same tide.

The Star of Greece made rapid progress south as she barrelled along before the monsoon winds. The usual calms around the southern approaches to the Laccadive Sea below Point de Galle saw the clipper delayed for several days until fitful southeast trades were fallen in with. Progress was slow as the ship sailed southeast towards a darkening sky. The sea became a foreboding grey-green as the wind began to strengthen and swing around to all points of the compass. Captain Shaw took readings from the barometer and the lead. Taking noon sightings or sightings from the stars was all but impossible as the drab and overcast skies continued to thicken and flashes of

19

Star of Greece

lightning outlined the approaching storm. The glass continued to fall rapidly as the crew worked frantically to shorten sail. The boats and anything moveable was double-lashed as Shaw set a course to run along the edge of the approaching cyclone.

The next few days proved sleepless for the crew as they fought to keep her ahead of the tempest. Captain Shaw ran his ship southeast by east away from the worst of the weather before turning south. Despite his fine seamanship, the edge of the maelstrom caught the ship. Sails were blown clean out of the bolt ropes as wind gusts of 60 to 80 knots and pouring rains battered the vessel. The cyclone blew itself out after three days during which time the Star of Greece had run southeast with the raging winds and seas coming off the port beam. Those aboard were thankful for Shaw's prescient navigation but also a little weary as he had perhaps carried more sail than was prudent. The clipper never the less battled her way successfully through the approaches to the Mozambique Channel on her way to Algoa Bay.

The favourable winds continued for the Star of Greece as she rounded the Cape of Good Hope and sailed up into the Atlantic. The trade winds proved fresh and reliable all the way home with Deal being sighted on the 18th of June, 105 days from Saugor. The Star of Scotia, on the other hand, had been delayed by the cyclone and then had to battle against frustrating headwinds. She did not make Deal until July 8th a full 20 days behind the Star of Greece. The Star of Greece passed customs inspection on the 19th and eventually made her way to the East India Dock to discharge her cargo of jute and tea.

Bill Shaw learnt upon his reporting to Corry's office in London that Robert Corry, the managing director and chief partner of JP Corry's was dead. Robert Corry had been a leading businessman in Belfast well noted for his charitable work and his involvement with the Harbour and Water Boards.

Star of Greece

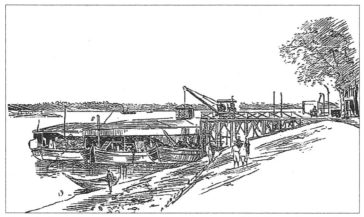

Jute hoisted by steam crane from barges which brought the jute in bales down the Hooghly from the interior.

Inside a jute mill.

Letters from India & Ceylon, Pg 46-9, 1895, John Leng & Co', Dundee.

Star of Greece

As a matter of principle, he had made a point of having all of his vessels manufactured by Harland & Wolff at their Belfast yard, to keep the money and jobs within the community. Corry's employed hundreds of people in all aspects of their business and with Robert's death, the direction of the company was open to question. However, with William Corry at the helm, the workers need not have worried. The timber and construction business was booming and the Star liners were bringing in regular and sometimes handsome returns.

The turnaround for the Star of Greece was swift, with her hold empty and her bilges cleaned out she was warped over to the dry dock for a shave and paint. Once cleaned Captain Shaw had his ship towed across to Bow Creek coal wharf (Near the East India Docks) to take on her load of coke. The design of the Star of Greece and her sisters was such that a fire in one part of her hold could not spread to the other two sections. The collision bulkhead would keep fire away from the business end of the vessel even if she collided bow on. The remaining twin bulkheads, one just abaft the foremast and the other just after the mid-deck house separated the usually highly combustible coal and jute cargos into three airtight compartments designed specifically to isolate and smother a fire that may break out within the stowed cargo. The Star of Greece was cleared out on the 18th of July and brought down to Gravesend the next morning. Shaw was forced to wait for the last of his crew and for the wind to come about so his ship could clear Dover.

Under these almost perfect sailing conditions, Shaw ordered the towline dropped abreast of the Deal light. Such were the vagaries of the light airs it took the best part of two days for the Star of Greece to make the Isle of Wight. The wind finally backed around to the northwest and freshened allowing the clipper to shake off her fog-shrouded lethargy. Conditions remained dull and humid as the Star of Greece passed the Scilly's on the 25th. From there she was 27 days to the Equator as the northeast trades blew hot and dry through the height of summer. Her usual run into the South Atlantic met with few obstacles as William Shaw piled on the canvas, determined to make a quick passage.

Star of Greece

All things went well until the Star of Greece ventured north out of the trade winds and into the doldrums south of Ceylon. A series of violent squalls and thunderstorms announced the early arrival of the winter monsoon. The winds picked up from the northeast as the clipper thrashed her way north up the Malabar Coast against the driving monsoon winds. The Hooghly was reached on October 18th, 1869, the Star of Greece setting a cracking pace. The ship had covered the distance from the Isle of Wight to Saugor in 87 days.

The stevedores wasted little time unloading the ship's cargo of coal into lighters before transhipping it into coal hulks and riverside bunkers. Once empty the apprentices and peggie's (ship's boys), cleaned out the bilge of the build-up of coal dust sludge and other noisome substances sloshing around in the dankest parts of the hold. The crew were set to placing old sailcloth and boards along the frames and stringers to prevent condensation from the hull contaminating the jute bales, to lower the risk of spontaneous combustion of the cargo. Loading of jute, hessian sacking, tea and hemp cordage was done in less than three weeks.

William Shaw had the Star of Greece cleared from Garden Reach on November 23rd. The stately clipper left her tug on the 25th and with the winter monsoon behind her the vessel bowled on down the Malabar Coast. Stunsails, skysails, watersails and a Jimmy Green were set to make the most of the prevailing northeast winds. For 22 days the leaden sky prevented accurate navigational readings as the regular monsoon squalls followed hard upon the heels of the racing clipper. By the 23rd day, the Star of Greece was well east of Mauritius. The wind was backing and veering from every quarter. The ocean became a dark grey, green and ominous clouds gathered overhead, all of which announced the approach of an out of season cyclone.

Captain Shaw kept a close eye upon the barometer as the atmospheric pressure plummeted. By the evening of December 14th the wind was blowing a howling gale, the clipper attempting to skirt around the worst of the weather. The ship rolled and dove continuously on a confused sea as enormous rollers dumped dollops of angry green water over the bulwarks and onto the deck of the Star of Greece. The storm raged for four days as the crew struggled to survive. The eye

Star of Greece

of the storm passed over the ship on December 17th and for a while, the clipper bobbed precariously about threatening to pitch her masts clean out of their steps. The wind soon shifted and intensified

There was a sharp crack from above as a violent squall caught the Star of Greece flat-a-back. As she heeled hard over almost onto her beam ends the fore, main and mizzen topgallant masts and yards came crashing down into the lower rigging. The ship's carpenter who was aloft attempting to cut loose the tangled rigging was hurled overboard into the raging sea and lost. The forestay and lower yards were badly damaged as the entire messed crashed down. Keeping her head to the wind Captain Shaw ordered both watches and the idlers aloft to clear away the wreckage. Axes, mauls and cold chisels were distributed from the bosun's locker and the crew set to with a will to clear the tangled mess before the topmasts and yards came down too. The clipper continued to run before the wind even as the desperate sailors began to set to rights the worst of the damage. All three topgallant masts along with the topgallant and royal yards had gone over the side, whilst the topmast rigging was in disarray. On deck things were little better with everything moveable having been swept overboard, boats stove in and other damage to ventilators, skylights and the like. After many hours of dangerous and back-breaking effort, the crew managed to clear the decks. The battered ship limped along as sailors laboured to set up a jury rig on the fore and main masts.

The Star of Greece was not the only vessel to take a hammering from the cyclone. The 1000 ton, 191 foot long, ship Darra, bound from Adelaide to London was partially dismasted, losing her foretopmast, main topgallant mast and bowsprit when she nose-dived into a gigantic wave; *"Description of narrow escape of fine ship Darra in cyclone from Cape Argus, 20 January:- arrived Table Bay 18 Jan. after fine passage of 59 days from Adelaide with cargo for London of 436 ingots silver, 7631 ingots copper, 3269 bags copper ore, 3920 bales and 29 packages wool, 1 pipe and 6 cases wine, 25 tons bark, 50 kegs preserved meat, 483 casks tallow, 13 bales leather and 9 passengers. Capt. Lodwick reports cyclone experienced 17 December 26.20S, 78E [Southern Ocean].*

Star of Greece

999-ton composite clipper Darra in port, circa 1875.
A.D. Edwardes Collection, State Library of South Australia.

Star of Greece

West India Dock, circa 1870s.
National Maritime Museum.

Star of Greece

*With barometer falling fast shortened sail and set hands to making
everything secure. Fore and main topsails blew away into the clouds.
Terrific gusts of wind. Lee main deck under water. Lee boat, hanging
under water, was cut away so as not to injure ship. Appeared danger
of everything being blown away. Sea made clean breach over ship
fore and aft, washing everything moveable from decks, including most
of livestock. In vortex of cyclone wind died away, but sea was
tremendous, tumbling over ship, washing right through saloon and
cabins. Ship then put under bare poles and allowed to drift. She
behaved herself nobly in the sea and, had she not been strongly built,
she must have gone down as at times sea was something frightful to
look at as it rolled down upon the ship."*

South Australian Chronicle, 16/04/1870

Other vessels suffered similar fates. The barque Crown broached
during the heavy weather and her decks were swept clean, she lost
sails and spars in the horrific conditions but managed to ride out the
tempest relatively intact. Other vessels in the area were not so
fortunate and more than one disappeared without a trace as the
cyclone passed out into the southern Indian Ocean.

The Star of Greece rolled on home as her crew continued to work up
jury-rigging and spars. Looking more like a stump topgallant rigged
vessel the clipper hummed along, making the line of Cape St Francis,
South Africa on January 9th, 1870, where despite her bald-headed
appearance asked to be reported "All well". The ship was assisted
around the Cape of Good Hope by light south-easterly winds, so
prevalent at this time of year. The ship eventually limped into
Jamestown Harbour, St Helena, having made her way west under a
limited jury rig of lower foresails & topsails. She dropped her hook
in the roadstead alongside dozens of other vessels many of which
sported damage from the cyclone and storms sweeping the Southern
Ocean. Safely at anchor rapid repairs were made with assistance from
the riggers and carpenters of Jamestown. With jury-rigged masts
fitted with topgallant yards and sails, the Star of Greece continued
north at a cracking pace as she fell in with the southeast trades. Under
this unusual setup, Bill Shaw guided his ship with skill and care

Star of Greece

passing the much damaged Darra 240 nautical miles southwest of the Cape Verde Islands, 86 days from Calcutta.

The winds north of here proved contrary. Shaw kept his vessel well out to sea searching for the prevailing westerlies that would drive his injured clipper home to London. This tricky navigational manoeuvre eventually paid off. Deal light and telegraph station was signalled on Thursday, March 17th 1870, notifying the Star of Greece's owners of her safe arrival. The pilot cutter came out that morning along with a steam tug offering to tow the ship into Gravesend. After some heated negotiations, William Shaw managed to get a price he was happy with. With line attached and a pilot aboard to guide the clipper into port, the Star of Greece reached safe anchorage at Gravesend, 112 days from Saugor. Having cleared customs and gained pratique the Star of Greece was brought up the Thames on the next flood tide. She passed through the West India Lock early on the 18th and was then warped into the wharf to discharge her cargo of jute, hemp and tea.

Whilst Captain Shaw was congratulated by William Corry for saving the ship, he was also cautioned about preventing further expensive damage to the brand-new clipper. Time was money and every day the ship sat idle was one when she was not earning her keep. The fitters and riggers worked feverishly to bring the vessel back up to A1 standard, a task they completed in little more than a week. Once more looking like new the Star of Greece was towed across to East India Dock to take on a fresh load of coke for Calcutta. The frenetic pace of loading cargo, preparing the vessel to set sail and rounding up a fresh crew was unrelenting.

With light winds from the east, the Star of Greece dropped the tug off Deal and with topsails sheeted home sailed down the Channel on a beautiful, glass calm sea. Pushing south by southwest the fresh and warm easterlies continued to the line which was crossed, on May 15th, 26 days out. The southeast trade winds proved reliable to the Cape of Good Hope and from there to Sandheads the summer monsoon winds pushed the Star of Greece rapidly up the Bay of Bengal in fine style. She hove to awaiting the pilot cutter off the mouth of the Hooghly on July 19th 1870, 91 days pilot to pilot. Upon arrival at Calcutta the Star of Greece took her place at the arrival buoys this time alongside her

Star of Greece

sister ship, the Star of Erin, commanded by Captain Jack Simpson. There was a friendly rivalry between Corry's skippers and with the arrival of the Star of Persia later in the month, the usual top hat was wagered as to which vessel would have the fastest return passage to London.

The Star of Erin had left Deal on April 3rd and arrived off of Saugor Island on July 9th, a run of 97 days, and the Star of Persia made the same run from London pilot to pilot. All three clippers were laid up in the Garden Reach waiting for the empty coal barges to appear. With their usual efficiency Calcutta's stevedores completed their work, different companies competing to load and unload the ships moored all along the river banks.   There was some delay in filling the hold of the Star of Greece, thus allowing her rivals the Star of Erin and Star of Persia to get underway well ahead of a frustrated Bill Shaw.   Captain Savage in command of the Star of Persia had his vessel loaded in the quickest time and set sail on August 1st. Not to be outdone Captain Simpson had his Star of Erin towed from the Garden Reach early the next day. An apoplectic Shaw could only stand on the poop and watch as the two Corry clippers disappeared down the Hooghly.   It would be another three weeks before there would be enough jute ready for stowing.

Corry's agents filled the 'tween deck spaces with chests of tea, bundles of hemp rope and bales of gunny sacking. August 19th saw the Star of Greece loaded down to her marks as the last bale of jute was screwed into place.   The pilot came aboard on the 20th and ordered the clipper to be warped out from her moorings and down to the departure buoys in the lower Garden Reach.   Stifling humidity and stench of Calcutta was left far behind on August 21st as the ship began the tedious yet dangerous 115 nautical mile trip to Saugor. Pilot and tug were dropped off on the 23rd as Shaw set to making up lost time.   He had a point to prove and with every tropical rag set 'alow and aloft', the Star of Greece cracked on south with a bone in her teeth and a hard-driving skipper in command.

The Stars of Erin and Persia were not having it all their own way. The prevailing southwest monsoon had forced the two clippers to run to the eastern side of the Bay of Bengal. As they raced south a series of

Star of Greece

violent storms battered them. The Star of Greece had a much better time of it, catching the monsoon at its height Bill Shaw followed the track of his predecessors but managed to avoid the worst of the storms as he swung his clipper southwest the trade winds coming abaft the port beam. Thirty days from Sandheads found the Star of Greece running south of Madagascar and headed for the coast of Africa. Shaw wanted to bring his vessel in as close as possible to the southern coast to take advantage of the land breezes and currents to be found there. As the clipper sailed towards Port Elizabeth the barometer fell steadily indicating an approaching cold front. The fiery pink sky of September 27th dawned as the Star of Greece approached the confused waters off Cape Agulhas.

The winds blew gale force from the west as the ship passed through the shallow and confused waters above the Agulhas Bank. Violent gusts and squall shredded sails as the crew hazarded life and limb aloft to shorten sail. The clipper laboured heavily with the wind gusting from four points, the lee rail constantly underwater and her deck awash. Just then a huge wave came over the bow causing the ship to dive, green water sluicing the deck the ship from just abaft the fo'c'sle to the break of the poop. The boats were lifted clean out of their blocks, and the front wall, skylight and doors of the deckhouse were smashed in. The contents were washed into the ocean through where the bulwarks had once been. Everything movable disappeared over the side.

The ship continually staggered under the weight of the water as she threatened to go under. True to her design she shook off the water, bobbing up like an iron cork. Hapless sailors in their wet weather gear struggled out of the scuppers and gutted wreckage of the deckhouse. The Star of Greece rolled on southwest into the teeth of the gale until she gained enough westing to clear the Cape. Smashing her way nor-northwest, the clipper under Captain Shaw's steady hand, clawed her way into calmer waters finally giving the exhausted crew some much-needed respite. The damage was quickly put to rights and new sails and rigging sent aloft. The thumping the ship took did little to slow her down.

Carrying jute to market, Calcutta circa 1870s.
Calcutta City Archives.

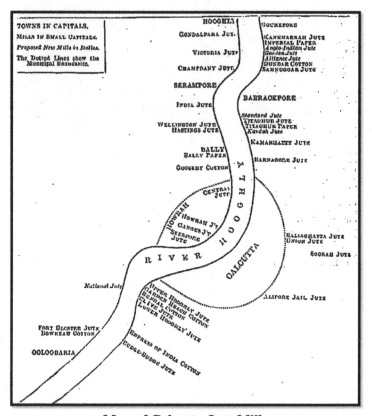

Map of Calcutta Jute Mills.
Letters from India & Ceylon, Pg 46, 1895, Sir John Leng, Dundee.

Star of Greece

Once clear of the westerly gales Captain Shaw again set all plain sail and the clipper cracked on up through the South Atlantic and into the more pleasant climes. The winds held steady, her royals never coming off her until just north of the equator when the doldrums left the clipper becalmed for several days.

The Star of Erin and Star of Persia encountered frustratingly light and variable winds for much of their journey north allowing the Star of Greece to make up ground. By the time the vessels of Captains Simpson and Savage were passing the Scillies the Star of Greece was snapping at their heels. The Star of Erin arrived off of Deal on the afternoon of November 14th, 107 days out, and took up her tow into the Thames that same night. Captain Simpson discovered that his ship was the first to arrive of the three and he perhaps felt certain that bragging rights for the fastest passage were his.

His confidence was short-lived as Captain Savage's Star of Persia reported her arrival off Deal the very next morning. His ship had also made the trip from Calcutta in 107 days. Both vessels arrived in good shape and were passed into their East India berths on the 15th and 16th of November, respectively. What neither man could have foreseen was William Shaw's reckless determination, uncanny luck and masterly navigation skills. For even as the Star of Persia was being shifted into her berth the Star of Greece was approaching Deal in tow. The battered ship passed Deal on the afternoon of November 15th, 84 days from Saugor. Upon tying up at his ship's East India berth, Captain Shaw and his much-relieved crew had cheese-eating grins planted upon their faces when they learned they had beaten the others by 23 days.

The cargos off all three ships turned a handsome profit for Corry's though they must have been wringing their hands in frustration at the damage Captain Shaw was causing to their new clipper. The badly mauled ship was placed into dry dock to be cleaned and painted whilst she underwent yet another refit to repair the damage sustained off of Cape Agulhas. A curious incident took place whilst Bill Shaw was waiting for his ship to be repaired. The Captain and his first officer were called to give evidence in a court case involving fraudulent dealings of a local merchant's assistant. Appearing before the Lord

Star of Greece

Mayor at Mansion House, London, a man named Walter Gunyon was charged with pocketing the proceeds of several purchases by various customers rather than passing the cash onto the business owner. Captain Shaw stated that he had handed over more than £20 to the accused and the first mate more than £4, none of which made it into the merchant's coffers. The case went to trial and the light-fingered assistant was convicted. The court case and repairs to the Star of Greece delayed her departure until the 9th of January.

The wind was blowing cold and fresh from the west southwest with accompanying snow squalls and hail. The barometer was falling rapidly announcing the approach of yet another cold front. Icy rain and overcast skies cast a dark pall over the English Channel as Captain Shaw and the pilot agreed to continue the clipper's tow down channel. They made it as far as the Isle of Wight on January 12th when it was decided that the wintery southwest gales were too much to fight against. The clipper and her tug put back to the Downs to wait out the worst of the blow. The Star of Greece dropped anchor on the morning of the 14th having battled the violent gales for 36 hours attempting to get back to a safe anchorage. The aborted effort to clear the Channel was not without its problems and several sails were carried away. It was another four days of running repairs before the ship was ready for sea.

"In westerly winds, the Downs are full of shipping outward bound, and waiting for a fair wind. Then on a dark night, the long line of their gleaming riding lights suggests to the spectator some great city in the sea...In easterly winds, the seaward-going host departs, and there come from the south and west the homeward-bound clippers, some in tow of steam-tugs for London, and others bound to northern ports, furling their sails for anchoring in the Downs till winds from the west and south spring up to bring them to their voyage end. The larger vessels anchor in the southern part of the Downs, in eight or ten fathoms of water, the bottom being chalk; while the smaller vessels bring up more towards the north, in the Little Downs, in from four to six fathoms of water, in splendid holding-ground of blue clay.

Star of Greece

Once an anchor gets into this blue clay, it will hold the vessel unless her chain-cable parts, or till she splits her hawse-pipes... and at last the sky clears in the north-east, and a golden haze enshrouds the fleet which on the waves lies heaving many a mile...on every ship the windlasses are manned. You hear the clicks of the palls as the anchors come up, and the creaking of the yards as they are being hoisted, and the singing of the sailors as they walk the capstan bars round, or heave the windlass handles... And then the sailors nimbly run aloft to loose the sails. The gaskets are cast off, the buntlines are let go, the clew lines hauled, and the great foretopsail bellies out before the freshening north-easter. Each ship spreads her wings, and they 'fly as a cloud and as the doves to their windows', presenting a wondrous spectacle of beauty from Deal Beach".

Rev T S Treanor of the Missions to Seamen, circa 1892.
www.eastkent.freeuk.com/deal/index.htm

With a favourable wind in the offing, the 157 ton, wooden paddle tug, Napoleon took the Star of Greece in tow from the Downs and out to the roadstead off Deal to await the expected wind change. The two vessels arrived on the afternoon of January 18th and dropped anchor for the night. The weather began to abate that evening. Mist shrouded the anchorage as dozens of vessels waited for the fog to clear. Captain Shaw kept a close eye upon the barometer which continued to rise steadily. Seas continued to be rough off of Dover but were slowly abating as the approaching high-pressure system moved in over the south of England. Conditions were much calmer the next morning with light and variable breezes.

The pilot and the Napoleon's skipper agreed that now was the time to get under weigh. With the anchor hauled short, a tow line was passed through the hawse pipe and as the tension was taken up by the tug the anchor cleared the water to be fished and catted by those upon the fo'c'sle head. Setting a headsail and reefed lower topsails, the Star of Greece, pulled along by the steamer Napoleon, once again headed out to sea. This time she managed to clear the channel pushed along by a strengthening north-easterly wind on a gentle, rolling sea. After an uneventful voyage, their arrival in the shallow, choppy waters of

Star of Greece

Sandheads was without major incident and the Star of Greece hove to on April 21st after 92 days at sea.

Jute was the lifeblood of Calcutta, having been grown there since time immemorial by Indian farmers who manufactured all manner of cloth, sacking and rope from its fibres. The soils around the city were of fine alluvial silt prone to frequent flooding, perfect for the growing of jute and hemp. The jute was produced in vast tracts of land by Indian farmers who were paid a pittance for their labours. It was a steady income and provided much-needed employment for the vast numbers of people flocking to the region. The raw jute travelled down the Ganges and its tributaries to the river port of Calcutta. From there, bundles of jute were transhipped by barges to the mills that dotted the landscape about the city. The mills then employed tens of thousands of labourers producing cloth, rope, gunny sacking and bales of fibre for export to British cities like Dundee which was a world-class centre of jute and hemp fibre weaving and rope manufacture. Jute clippers and steamers large and small anchored below Hooghly Bridge near the railway wharf where their loads of coal were discharged. Small barges and riverboats would then arrive from upstream with loads of jute and hemp which would then be transhipped into the holds of the waiting vessels.

"The Hooghly resembles the Clyde below Glasgow Bridge or the Thames below London Bridge. The only bridge between Calcutta and Howrah is a floating bridge, supported on boats strongly moored, with two movable portions, which swing open for a certain length of time twice a week to allow large craft to pass through. During the busy hours of the day, the Bridge traffic presents an animated scene, from the curious teams and still more curious features and dresses of the pedestrians. Below the Bridge on the Calcutta side are first the large steamers moored to the quays belonging to the great passenger and Trading Companies. Then in long tiers come the large fleets of the British India and other Steam Navigation Companies, whose steamers sail to all the ports in the East. The British India alone is said to have a larger number of vessels, and also a heavier tonnage, than the great Peninsular and Oriental line. After these come the big tramp steamers, and the long rows, three, four, and even more tiers

Star of Greece

deep, of three and four-masted sailing ships, with many of whom we are familiar as jute-laden vessels coming to Dundee.

Finer ships in the commercial marine are nowhere to be seen than those engaged in the jute exporting trade. Two or three steam yachts, the Viceroy's amongst the number, and several gunboats, despatch boats, and guard ships complete several miles of shipping lying in the river... The business part of Calcutta, chiefly connected with the Jute Trade, is known by the natives as Alloe Godown, signifying Potato Warehouse, for which the ground was formerly used. Now it is covered by large and commodious offices, the upper floors of the building being generally occupied for residential purposes by the merchants or their European assistants. At the doors are generally a number of gharries, or street cabs, differing from horse cabs in being only wide enough for one person on each seat, and having sliding doors and louver openings instead of glass windows. Each office keeps two or three gharries of its own, as nobody thinks of walking even a hundred yards. Every jute broker has his gharri (horse-drawn carriage) to facilitate his movements from office to office or to the jute bazaar."
Letters from India & Ceylon, Pg 48-9, 1895, JOHN LENG & CO, Dundee.

The discharge of the Star of Greece's coal cargo was initiated almost immediately with lighters hove to alongside as the crew cockbilled the lower yards for use as cranes. The stevedores or coolies as they were known colloquially worked for 12 or more hours a day for little pay dressed in cotton loin wraps in the stifling heat. The coal was lifted out in large wicker baskets to be dumped right into one of the many lighters lined up to take the coal to the railway yard bunkers. With her holds relieved of such a heavy burden it was only the 120 tons of kentledge in her limbers that kept the clipper upright. This period of riding in the water high above her marks did not last long as Corry's Calcutta agents had managed to secure a jute charter at £3 per ton, equal to 8500 bales, worth around £5400 to the company. Thus taking out wages, supplies and running costs, the Star of Greece and her sister the Star of Persia had already more than paid for their construction. It was a point of contention between J.P. Corry's and their captains that unnecessary damage was eating into

Shipping on the Hooghly, Calcutta, circa 1870s.
Johnston & Hoffman Postcards, Calcutta.

Star of Greece

Coolies loading coal at Kidderpore Coaling Dock, just downstream from the Hooghly Bridge, Calcutta. Circa 1890's.
Old Indian Photos.

Star of Greece

the company's burgeoning profits. However, with freight rates varying between a paltry £2 3s 6d and £5 per ton Corry's were never going to lose out even if the ship and her crew were lost.

The jute was loaded in the quickest time so far and the ship was cleared out from her loading berth on May 7th, a fortnight after the clipper's arrival. The summer monsoon had arrived on time and Captain Shaw was forced to take the Star of Greece south by southwest across the Bay of Bengal. The ship cruised along at rapid pace as she fell in with the southeast trades. Algoa Bay was approached some 30 days from Calcutta and Captain Shaw set course to cross the Agulhas Banks. Confusing cross-currents and heaving seas were encountered as the north-easterly winds drove the clipper southwest around Africa. Weak westerly winds were encountered as the ship rounded the Cape of Good Hope for a favourable slant into the South Atlantic.

Several of the crew were laid low with dysentery, possibly from tainted salted meat or fouled water in the tanks. With fewer crew available to set and take in sail, Bill Shaw was forced to run under shortened canvas or risk his ship becoming unmanageable. He worked hard to cure his afflicted sailors using the numbered bottles from his medicine chest. Much of the salt junk was a putrid greasy mass resembling soggy brown paper pulp, and the ship's biscuits were flinty, weevil filled and rock hard. For whatever reason the crew were sick and the ship was deep in the South Atlantic far from qualified medical help. Shaw could have put into St Helena but the island was to windward and there would be little chance of reaching it before one or more of the crew expired. Despite the Captain's most attentive ministrations 38-year-old Able Seaman Alexander McNeil, died from the effects of dysentery. A traditional sea burial took place after a brief service conducted by the Captain attended by all except the men at the wheel, the lookout and those too sick to leave their bunks. The body with great care and reverence was sown into a canvas sack made by the sailmaker and weighed down with iron. Laid upon a plank of wood and covered by a red merchant flag, McNeil was consigned to the deep.

Star of Greece

Having rounded the Cape, the Star of Greece ran headlong into a series of north-westerly gales running north up the Skeleton Coast. The ship and her crew were more than capable of handling such conditions as hurricane-force winds and huge seas swept down. Head reaching into a howling north-wester the clipper regularly took seas over the lee rail with many of the slower crew having to be lifted from the scuppers. Even with lifelines and netting strung along the sides, there was always the chance of being washed overboard. Such a sea climbed over the topgallant rail and washed the luckless second officer, 21-year-old Edward James Grounds, clean over the side. Grounds was a young apprentice just out of his time, and out of luck. The weather conditions were so severe that there was no thought that a boat could be successfully launched to attempt a search for the lost sailor. To do so would have probably resulted in many more deaths, and so the Star of Greece crashed northwards in search of milder climes.

The run home was without further serious incident and approaching the Scilly's a series of violent thunder squalls from the east caused Captain Shaw some concern. Winds were light and conditions cloudy as the storms were left behind. The Star of Greece entered the English Channel on August 16th. The wind veered round to the southwest allowing Shaw to sail his ship all the way to Deal without having to engage a pilot. He ordered the ship hove to, within sight of Deal pilot station on the afternoon of August 18th, waiting for the pilot cutter to approach. The voyage from Saugor to Deal had taken 98 days and cost the lives of two sailors. The trip turned a sound profit but was not without cost. The Star of Persia had made the run from Calcutta in 102 days, another victory for Captain Shaw, but not one he and the crew felt like celebrating.

Despite the continued setbacks, the rapid turnaround of the jute clippers at the East India Docks had become routine. They warped into the same berths, were unloaded by the same stevedoring company and were refilled with their regular loads of coal. The whole process had become something of a production line for Corry's and their captains. The Star of Greece was soon ready for sea with a fresh crew on September 20th after a clean and refit. The weather was unsettled with thundery squalls and storms sweeping in from the

Star of Greece

northwest. The clipper was towed from Gravesend to Goodwin Sands on the 23rd. By now the wind had backed around to the west-northwest and abated making departure possible. Visibility was poor and conditions hazy as the tug hauled the Star of Greece out into the Channel.

With the wind from the north, the ship sailed down the Channel at a good clip. The choppy conditions continued until they were just south of the Scillies having made excellent time so far. The glass was dropping sharply as the wind shifted yet again, this time blowing a howling gale from the west southwest and whipping up short high crested waves which broke over the bow of the Star of Greece in rapid succession. She was in a precarious position in the heart of the Bay of Biscay when the worst of the weather hit. Captain Shaw was forced to beat down across the Bay into the teeth of hurricane winds, mountainous, white-capped swells and driving rains. The Bay of Biscay had the worst weather in the North Atlantic and was a regular graveyard for ships and their crews. The Star of Greece did not escape unscathed as a comber came rolling over the rail and carried away most of the crew upon the deck. When the count was taken, the sailmaker, 25-year-old Lindsay Robinson had gone over the side with no chance of rescue. His was the third death of the voyage.

Despite the horrendous start to the voyage, the weather improved rapidly sending the Star of Greece romping south. The Cape Verde Islands were sighted on October 6th, and the equator crossed on October 24th, 1871, 30 days from Deal. Having successfully negotiated the stormy waters off the Cape of Good Hope the crew spent Christmas Day celebrating with a 'make and mend; day aboard the ship as she beat up the coast towards the Hooghly. The going was slow as the clipper worked her way north against the prevailing monsoon winds. The muddy waters of the Ganges Delta soon appeared signalling that the journey would soon be at an end. The Star of Greece dropped anchor in the Saugor Roads on December 28th, 95 Days from Deal.

As the ship was being warped into the mooring buoys, her sister the Star of Persia was being towed downriver, ready to put to sea. The coal was quickly discharged opposite Kidderpore Coaling Dock.

Star of Greece

Turnaround time was just two weeks, enough to resupply and take on two new crew to replace those lost at sea. Every load of coal coming into Calcutta had to be taken off into lighters on the bank opposite Kidderpore. The coaling docks were not long enough nor the water deep enough to accommodate all the ships demanding coke for their boilers. The Kidderpore side was also the major terminus for the Calcutta Railway service operated by the Port Railway Trust. The jute traffic was concentrated in the Sealdah, Howrah and Cossipore Goods Yard. Most jute was transported by rail or barge downriver from the Kidderpore docks to be loaded aboard waiting clippers and steamers.

The stevedores of Calcutta were as good as their reputation, the ship's hold, in short order was filled with rice, cotton, hemp and jute. Loading was completed in record time and Captain Shaw had his clipper cleared out on January 12th and taken across to Garden Reach for final departure preparations. As this was happening the Star of Erin was being towed into the coaling station to discharge. The side-wheeler tug that would tow the Star of Greece the 115 miles to Saugor arrived on the morning of the 14th of January 1872. Once in the Saugor Roads, the Star of Greece sailed gently down to Sandheads anchorage and hove to, allowing the pilot schooner to pick up its charge. Shaw then ordered topsails and headsails set to clear the outlying sandbars and shoals. With a quartering wind, Bill Shaw was confident that his ship would make excellent time.

The Star of Greece sailed headlong into one of the worst periods of cyclonic weather on record. The barometer rose and fell dramatically over the several weeks that it took to sail from Calcutta to Cape Agulhas. The first of the storms struck as the clipper passed below 10° south. Sailing southwest Captain Shaw bore his ship away to the south to avoid the worst of the weather. Hammering along on a westerly course the Star of Greece was forced to run the gauntlet of a serious of vicious squalls and a cyclone as she slid south of Madagascar and into the lee of the island. Shaw kept sail on for longer than was prudent, as was his way and managed to get his ship through with just the usual blown sails, swept decks and strained rigging. Once clear of the Cape of Good Hope, the clipper flew along into much calmer waters as the southeast trades were picked up earlier

Star of Greece

than would normally have been the case. They proved fitful and none too reliable. A flying start was followed by a slow finish passing Deal on April 19th, 93 days from Sandheads.

East India Dock was a hive of activity for Corry ships. Moored alongside the Star of Greece was her rival the Star of Persia and the 1000 ton Star of Albion, all loading coal for Calcutta. Captain Shaw was currently considered to be Corry's crack clipper master and John Simpson of the Star of Persia was determined to lower Bill Shaw's colours. Captain Hughes of the Star of Albion was equally ambitious, his vessel had yet to match the passage times set by the larger ships. The Star of Greece was taken into dry-dock for a clean and to have her rigging repaired after her dusting in the Indian Ocean. The first clipper away was the Star of Persia which cleared out from Gravesend on May 3rd and passed Deal the following day. Captain Hughes managed to get the Star of Albion down to the Downs on May 11th sailing from Deal the same afternoon.

Despite a slower start than his rival, Shaw was confident that the performance of his ship would eclipse his rivals, after all, there were reputations at stake. The Star of Greece cleared out from the East India Dock on May 20th and was towed to her Gravesend mooring the next day. The clipper picked up her tow on the high tide of the 21st and was taken out to the Downs roadstead to await a favourable slant on the wind. This came the following morning when a moderate breeze sprang up from the northwest.

Under topsails and headsails, the Star of Greece cracked on to the north coast of Africa in search of the prevalent northeast trades. The ship crossed the equator along the line of 24° west on June 16th, 25 days out, and if the winds continued favourable, a record run was in the offing. The Saugor light was sighted on August 10th after just 80 days at sea. The Star of Greece had rolled past the vessels; British Envoy, Daylight, Monto Rosa, Lord Strathnairn and Delaware, on her way to Calcutta.

Star of Greece

The British Envoy, 1265 tons, Built 1866, T. Royden and Sons, Liverpool.
A.D. Edwards, State Library of South Australia.

Star of Greece

Garden Reach at Calcutta, circa 1865.
Vibart Collection, British Library.
She tied up at her Garden Reach mooring on August 12th and William
Shaw was certain that his ship had recorded the quickest time of the

Star of Greece

three Star Liners that had left London in May. Captain and crew were to be sorely disappointed. They discovered that the Star of Persia had set a world record for the run from the Cape of Good Hope to Calcutta, 22 days, the fastest time for any vessel, sail or steam ever made to that date. The Star of Persia had arrived at Saugor after 73 days from Deal, her entire run from Gravesend to Calcutta occupying 77 days. The Star of Albion performed admirably but was outmatched by Corry's crack clippers. Captain Hughes and his ship were towed into Garden Reach on August 17th after what would normally have been a creditable voyage of 98 days from Deal. The race for overall voyage length was on as Captains Simpson and Shaw went to extraordinary lengths, pushing the stevedores and their ships' crews to the limit to ensure that their vessels would be loaded and ready for sea in the quickest time possible.

Having been first to arrive the Star of Persia was first to get away, slipping her moorings on the Hooghly on August 9th 1872. The Star of Greece was loaded and ready to sail after just 12 days in port and departed her tug at Sandheads on August 27th. Her hold filled with cotton, rice, jute and linseed made for a rather crank ship and extra ballast was needed. The luckless Captain Hughes discovered that Shaw and Simpson had snaffled much of the available jute for their own vessels holds. Thus his clipper, the Star of Albion was still loading in early September. She finally set out from Saugor on September 8th almost a month behind the Star of Persia.

All three ships were forced to take the wide track skirting the eastern side of the Bay of Bengal as the southwest monsoon blew strong and steady. The easterly set of the current meant that the clippers had to sail well south into the area of the southeast trades before they could make westing. The Star of Denmark, which had departed Calcutta much earlier on July 25th had run the same course as the current three Corry clippers and had ploughed headlong into a gale east of Madagascar, on August 19th. The crew of the Denmark discovered that she had sprung a severe leak and were forced to man the pumps continuously and throw 250 tons of cargo overboard. The leak was eventually found and plugged but it took until September 23rd for the vessel to drop anchor at Port Elizabeth for repairs. Aboard the Star of Greece the usual monsoon squalls accompanied the clipper south

Star of Greece

and one luckless crewman, 21-year-old James Kent, who had signed on despite being grievously ill, succumbed to Tuberculosis on September 14th.

The Star of Persia took line honours reporting her arrival as she passed St Catherine's Point on November 19th, 100 days out. Despite being held up by contrary winds off the Isle of Wight John Simpson's clipper had set yet another record. Her round trip from Gravesend to Calcutta and back had taken a total of 6 months and 12 days. After the checking of records, this proved to be the fastest round trip by a sailing ship ever on the Calcutta run. Unbeknownst to Captain Simpson, the Star of Greece was closing fast upon England. The clipper had run into a series of 'dead muzzlers' (headwinds) and almost non-existent trades as they meandered north. Bill Shaw's crew let go the anchor off Folkstone on December 4th 1872, 99 days from Sandheads. After such a dismal run William Shaw was perhaps hoping that the Star of Persia had also struck fickle winds on the voyage home.

It was not until two days later when the Star of Greece dropped anchor at Gravesend that he discovered how close the race had been. The Star of Greece's total voyage time was 6 months and 16 days, close to a record but not fast enough to beat her sister. To celebrate his victory John Simpson had a brass gamecock cast to be placed high upon the truck of the mainmast of the Star of Persia announcing to the world that his ship was 'Cock of the Route'. The fastest clipper on the Calcutta run. Rival captains took it as a challenge, chief amongst them, William Shaw. Bringing up the rear as 'Tail-end-Charlie' came the Star of Albion, skirting by Deal on December 12th. Captain Hughes had pushed his little clipper to the limit and managed to make the quickest passage of the trio, a run of 95 days from Saugor. So even without the Blue Riband, he could still claim bragging rights for the fastest journey of Corry's clippers home that year.

Despite William Shaw's initial disappointment, he did not let it distract him from the business at hand. The frenetic pace of commerce continued with the Star of Greece being entered into the East India Dock for loading on December 6th, just two days after her arrival. The discharge of her cargo of cotton, rice, jute and sacking

Star of Greece

was completed over the course of 10 to 12 days before the clipper was moved across to the coaling staithes. Whilst this was going on the Star of Persia set forth from Gravesend on December 27th bound again for Calcutta, John Simpson hopeful that his ship would improve on her previous record-breaking round trip. The crew of the Star of Greece spent the Christmas and New Year holidays ashore waiting for the final clearance of their ship. She was entered out of the London Docks on January 6th and brought to Gravesend to await the last of her crew to come aboard.

The Star of Greece continued to lay at anchor through January 13th as the south-westerlies brought with them periods of snow, frosty nights, thunder-squalls and fog. The Downs and Deal roadsteads were becoming exceedingly crowded with sailing ships looking to depart the moment the wind shifted. In the end, Shaw and the pilot decided that with the passing of the latest cold front a drop off in the wind would allow the Star of Greece to be towed to more open waters.

Once clear of Cape de La Hogue the clipper dropped her tow and made westing out into the North Atlantic. With force 5 winds from the southwest accompanied by driving rain, the Star of Greece was well out into the Western Ocean before the captain altered course sailing due south. The Star of Greece made excellent time reaching the equator on February 14th, 32 days out. There she lay becalmed in company with the clipper Parsee bound for New Zealand. The two vessels were in each other's company for 4 days before the southeast trades kicked in and the two parted company.

One crew member, 36-year-old Able Seaman, Sam Scholer, died of a heart attack on February 23rd as the ship was crossing the South Atlantic. The final leg of the voyage, around Cape Agulhas and up through the Indian Ocean was completed in relatively good time. The Saugor anchorage was reached on April 19th, 96 days from Deal. There was a three-week layover waiting for the season's jute to be loaded. Cleared out on May 15th, the crew of the Star of Greece were happily surprised to witness the arrival of their battered fellow the Star of Denmark. It was feared that she may have foundered as the ship had sailed from Gravesend on January 5th. Her passage of 120 days saw the vessel arrive in a rather sorry condition. This at least

Star of Greece

solved the mystery of the Star of Denmark's apparent disappearance and brought relief to her owners back home.

The summer monsoon had arrived right on time. Upon departure from the Hooghly Captain Shaw sailed his usual route down the west side of the Bay of Bengal to clear the reefs and shoals that were to be found west of the Andaman Islands. Here the Star of Greece ran afoul of the areas frequent squally weather, and heavy rains. As she approached the area north of the equator the clipper encountered and frustrating time of light, unsteady winds punctuated by showery weather and frequent thunder squalls. The lightning filled tempests did not begin to abate until the ship was well into the southeast trades. Once there the Star of Greece rolled on west-southwest towards Africa. Upon approaching the Cape the clipper was held up by strong to gale force westerlies that forced Shaw to tack against the prevailing winds. There was considerable damage to sails and rigging as the crew struggled to keep the ship on course.

With supplies running short Captain Shaw decided to stop in at the island of St Helena to top up his water and sea stores. The Star of Greece arrived on August 19th 1873, 63 days from Saugor. The clipper anchored off Jamestown and her captain sent a boat into the harbour to purchase supplies and to report their arrival. The short respite gave time for the crew time to carry out some much-needed repairs and gain some sleep as both watches moved to day work hours, 6:00 am to 6:00 pm. The layover was short-lived with Bill Shaw anxious to get back to London. The Star of Greece set sail from St Helena on July 21st, and onto the Cape Verde Islands. She passed Lizard on August 27th. It was off of the Isle of Wight the next morning that they picked up a tow, passing Deal later that day. Northwest to southwest gales made travel up the Thames difficult and the Star of Greece was forced to drop anchor in the Downs until winds abated on the evening of the 29th. The clipper did not make Gravesend until August 31st, thus ending a long and arduous 108-day journey.

The Star of Greece was sent to the dry dock for a refit and clean, a semi-annual event that enabled the ship to maintain her record-setting passages. The ship cleared out from London on October 1st being towed to her Gravesend mooring the same day. Captain Shaw had

Star of Greece

his ship shifted out past the Downs the next morning and with a fresh northerly breeze behind her the Star of Greece romped on down the Channel and out into the Bay. The wind backed around to the southwest as Shaw steered his ship well west of the Bay of Biscay to avoid the worst of the cyclonic weather brewing up there. Cape Verde was passed on October 9th as the Star of Greece galloped south.

On October 25th a violent squall struck the ship side on and a particularly nasty wave swept over the ship knocking several sailors into the scuppers and smashing bulwarks. There was the cry of 'Man overboard!' as Able Seam William Mulligan was washed over the side and out into heaving seas. There was little chance of rescue in such treacherous conditions and thus with regret, Bill Shaw pushed his ship and crew onwards. The run around the Cape of Good Hope and up into the Indian Ocean was without major incident. The northeast monsoon made sailing into the Bay of Bengal a rather tedious affair. The Star of Greece was brought into her Calcutta moorings on the 7th, 98 days from London. The Star of Persia was already in port, John Simpson having left London on September 15th, bringing her out to Calcutta in 115 days from the East India Docks.

John Simpson and William Shaw, the two great rivals, and friends exchanged commands for the voyage home. The Star of Persia had suffered storm damage on her way out to Bengal and William Shaw was laid up with fever. The Star of Greece was ready for sea before her skipper was well enough to travel and so the two men agreed that John Simpson would take her home whilst Shaw would sail home in charge of the Star of Persia once both were ready for sea. The Star of Greece departed Calcutta on February 2nd. The Star of Persia was in dry-dock undergoing a refit as William Shaw lay abed recuperating. The crew of the Star of Greece said farewell to the pilot on the morning of February 4th hove to in the Sandhead Roads, clearing the bar on the rising tide. The prevailing northeast monsoon allowed Captain Simpson to let the clipper have her head.

With every stitch of canvas aloft the Star of Greece raced across the Bay of Bengal and down the Coromandel Coast. She passed through the usual period of calms west of Ceylon and sailed into the Indian Ocean. Right around into the South Atlantic, the clipper was met by

Star of Greece

weeks of light and contrary winds and for much of the time, the trade winds failed to appear. The fickle airs continued to Cape Verde when the Star of Greece ghosted past at the beginning of May. Light and variable northeast to northwest winds and smooth seas were encountered for much of the trip north. The Scillies were passed on May 18th and the Deal anchorage reached on May 20th 1874, 104 days from Sandheads.

As the Star of Greece was coming to anchor off Deal another Star clipper, the Star of Scotia, commanded by Captain Edward Hughes, arrived in Hobson's Bay, Victoria with 26 migrants and a hold filled with general merchandise. The ship had departed London on February 22nd and arrived at Melbourne on May 20th, 87 days out. Whilst the Star of Greece was wallowing in the calms of the South Atlantic William Shaw was getting his temporary command underway. The Star of Persia sailed from Calcutta in the last week of February at the height of the monsoon season. The tailwinds were a blessing that allowed a quick passage southwest. Driving monsoon squalls were the norm as the clipper raced across the Indian Ocean towards Madagascar.

Unlike the Star of Greece, the Star of Persia after her refit was speeding home at a rate of knots. With the hard-driving Bill Shaw in command, she passed Ascension Island on April 15th and raced north with the southeast trades at her stern. Shaw drove right up the English Channel passing Deal on the evening of May 28th before heaving-to in the Downs just after midnight the next day. Both vessels were soon tied up at their East India berths where the two captains met to compare notes.

The Star of Greece was almost ready to depart when the Star of Persia arrived. Bill Shaw had still not completely recovered and so the Star of Greece did not clear out from London until June 30th 1874. The clipper departed Poplar on July 1st for Gravesend and passed Deal early on the 2nd. The Star of Persia with James Mahood in command was hot on her heels, sailing from London on July 7th. The two vessels upon approaching the equator were in constant company and there was rarely more than a few hours between them as they raced their way to Calcutta. The twin clippers entered the Bay of Bengal at

Star of Greece

the end of September pushed along by the southwest monsoon. Approaching the Hooghly the weather was fine, with passing clouds fillings the sky as localised showers fell across the Bay.

The Star of Greece hove-to off of Saugor pilot station on October 2nd, 92 days from Deal. She then proceeded up the Hooghly escorted by her sidewheel tug. Less than 24 hours later the Star of Persia appeared at the anchorage, thus proving that for this voyage at least James Mahood had the faster vessel with the fine run of 86 days from Deal to the Saugor roads. The Star of Greece arrived in Calcutta on the 4th with the Star of Persia tying up alongside the next morning. The Star of Persia still carried the brass cock at her masthead, and it was a trophy that William Shaw was determined to make his own.

The two clippers had arrived in Calcutta none too soon for down south in the Bay of Bengal the signs of an approaching intense tropical low were in the offing. On the morning of October 14th the 1150 ton ship Ireshope, 250 miles south of Calcutta, experienced increasing squalls with winds from the north-east. By early evening the ferocity of the squalls increased dramatically with winds strengthening to gale force from the north. At midnight the wind veered round from north to north-west as a terrific squall slammed into the ship shredding what storm sails she still had aloft. The horrendous might of the cyclone was felt on the morning of October 15th when it crashed into the coast on the western side of the Hooghly mouth. The wind blew with hurricane force from the early morning, as seas rose rapidly. The barometer fell from 29.68 at midnight to 28.94 by 5:00 a.m. The tempest was then raging with increased in fury until 9:00 a.m. when the barometer had fallen to 28.15, after which the preternatural calm of the storm's eye passed over. The eerie lull lasted all of 15 minutes after which the storm proceeded to blow with greater force and violence than before.

The wind continued to blow at until early evening when the barometer began to rise indicating that the worst of the weather had passed. The cyclone continued to smash its way northwest for another two days. Vessels caught at the head of the Bay, in the Saugor roads, or the lower Hooghly were helpless against the storms fury and many smaller craft foundered, their crews drowned. The water at Diamond

Star of Greece

Harbour at 7:00 p.m. on the 15th, an hour before low water, was 16 feet above the level at the same hour of the day before and only 3 feet above the previous high tide, thus negating much of the potential flood damage that would have come from a tidal surge up the river and into Calcutta proper. The storm ravaged the area in and around the city of Midnapore. There was little to no flooding but such was the fiendish nature of the wind that the final death toll was close to 80,000 people. Calcutta escaped catastrophe this time and the crews of the twin Corry ships had the chance to witness the damage caused to shipping as vessels limped into Garden Reach. The 900 ton barque Evening Star of Bombay and the ship Grand Duchess of Liverpool (195 feet long 1300 tons) were not so lucky, the vessels being wrecked. From the Evening Star's crew, just three men were saved, whilst those aboard the Grand Duchess all drowned in the storm-ravaged waters off Saugor Island.

Captain Shaw had his ship loaded and ready to go by October 20th, departing three days hence. The crew were witness to bone-chilling sights as they were towed down the Hooghly. The river was choked with the detritus caused by the cyclone. Amongst the wreckage were hundreds of bodies of animals and people being swept out to sea on the outgoing tide. The nauseating stench of putrefaction and death was heavy on the air. The sandbars and shoals of the Sandheads were littered with wreckage from lost vessels and the waters of the delta were blackened from the filth washed downriver. The Star of Greece set sail from Saugor on October 23rd sailing out into the debris-filled waters of the Bay of Bengal.

The Star of Persia began her run from Saugor on October 29th, 1874, with Captain Mahood determined to catch his compatriot on the run home. The Star of Greece had barely cleared the outer shoals of the Hooghly when she was struck by tragedy. In the generally smooth waters at the head of the Bay an experienced sailor, 46-year-old Thomas Galloway was aloft with the other members of his watch setting sail when he lost his balance and plummeted to the deck below. The grief-stricken crew could do nothing for the unfortunate man as he lay dying from his injuries. The deceased was buried at sea once the clipper reached the deeper waters and the crew returned to their assigned tasks with little time to brood over their shipmate's sudden death.

As the Star of Greece was homeward bound, the Star of Erin which had departed Belfast on December 1st bound for Calcutta, encountered some wild weather of her own. As she crossed the Bay of Biscay on December 10th 1874 the ship was met by thick and dirty conditions. By the morning of the 11th the crew of the clipper were battling for their lives, and by the 12th hurricane winds had the vessel hard over to leeward with high cross seas breaking on board fore and aft. At 3:30 pm a large wave broke right along the ship and stove in the boats on the windward side. The winds increased even further as the afternoon wore on and the vessels' captain was informed that the mainmast had been sprung at the step and threatened to roll right out of the ship. The crew were ordered to tighten the stays as the rhythmic swaying of the masts increased with the roll of the ship. The spars above whipped about in their slings threatening to come crashing down upon the decks. Captain Mills, fearing that the mast butt was about to drive itself through the bottom of the ship, ordered the crew aloft to cut away the main topmast.

Equipped with axes and knives the sailors quickly accomplished this task with the result that the royal, topgallant and upper topsails yards were carried away, also taking out the fore topgallant mast, mizzen top and spars. The mainmast continued to sway violently in its step. The aft-most mast snapped off a few feet above the deck and disappeared with the remaining spars over the lee side of the ship. With little more than a stumped mainmast and a missing mizzen, the Star of Erin continued to roll heavily in the steepening swells. The captain ordered both bow anchors let go to keep her head to the wind. There the crippled clipper stayed until the wind and seas abated enough for the crew to set up a jury rig. Through careful navigation and seamanship, Captain Mills and his crew guided the Star of Erin into the safe anchorage of Ria de Vigo on Spain's northwest coast. Once in the sheltered waters of the river a tug was engaged to bring the stricken ship into Vigo's harbour. Stripped of her damaged gear the vessel was towed round to the Belfast shipyards of Harland & Wolff for a complete refit.

Star of Greece

The Star of Greece and Star of Persia raced home to London. Despite being no more than a week behind Shaw's clipper, Captain Mahood's ship was unable to gain ground. On January 15th 140 miles southwest of Ushant, cracking along ahead of a howling sou'wester, the Star of Greece passed an abandoned Norwegian built wreck. The mysterious vessel had nothing standing apart from her mizzen mast stump, her deck submerged, bulwarks washed away. The crew could see no name for the derelict, she was yet another victim of the violent gales that were sweeping the Bay of Biscay. The Star of Greece passed into the Downs on January 16th, 83 days Saugor to Deal.

Poplar came into view on the 18th as the crew were making ready to warp the ship into her regular East India berth. The Star of Persia signalled her arrival off Deal on January 29th 92 days out and passed through the lock into the East India Docks on January 31st alongside her sister. The Star of Erin's poor run of luck continued, for as she was being towed into Belfast on February 6th, the tow rope parted in the rough conditions and the clipper ran headlong onto a mud bank. Two tugs were immediately sent out to pull her free but to no avail. The ship was stuck fast and not refloated until the rising of the tide 24 hours later. From here the Star of Erin was manoeuvred into the Queens Island Dock that afternoon.

Captain Shaw needed to oversee repairs to his ship as she had suffered damage to rigging, spars and sails as she rolled home in the winter storms of early January 1875. The Star of Greece was laid up for a month. The last week of February saw the ship warped across to her coaling berth for loading. The clipper was cleared out on March 3rd and shifted down the Thames on the 5th. With the wind from the northeast Bill Shaw was able to drop the tow-line off Deal and sail down the Channel under all plain sail. Seas were smooth until west of the Scillies when conditions became rough as an easterly breeze whipped up the shallow waters in the Bay of Biscay. Shaw pushed his ship onwards, making Cape Agulhas in 48 days. Here they encountered the 431-ton, Sunderland barque Salome, bound from London to Brisbane. The Star of Greece was making her way around the bottom of Africa along the line of 38° south in fine weather. The two vessels stayed in each other's company until William Shaw

Star of Greece

ordered a course change to the northeast in search of the south-easterly trade winds.

The Star of Greece made the Saugor roadstead on May 23rd, after just 78 days at sea, the fastest ever run by the Star of Greece to date. Captain Shaw and his crew were in high spirits as the ship meandered up the Hooghly for there was every chance that the Star of Persia's record was up for the taking. Her time in Calcutta was just 21 days. She took her departure from Garden Reach on June 15th. The Star of Greece spoke to the barque Gladiolus of North Shields as the vessel was bound south, on August 21st, 1875, 65 days out and passed by Cape Verde two days later. Shaw determined to break the Star of Persia's round trip record, pushed on arriving off Prawle Point on September 20th his vessel having struck light airs and headwinds on her way to the Channel. The clipper's voyage time from London to Calcutta and back was 6 months and 18 days, the same time to the day as the year before.

The trade in jute was booming for Corry's and they had already invested in three much larger jute clippers; the 1337-ton Star of Germany in 1872, the 1981-ton Star of Russia and the 1644-ton Star of Italy, both launched in 1874. The company's ships were rarely in port for more than a month unless undergoing repairs or survey. The Star of Greece was taken in for regular work at the end of her latest trip before being sent across to her coaling berth. The ship was loaded and cleared out from Poplar on October 31st before being taken down to Gravesend on November 3rd 1875. Her hull was clean, her rigging tight and the ship in near-perfect trim. Captain Shaw was well rested and eager to be away. The tug was greeted the following day which hauled the Star of Greece out to the Downs roadstead to await her final clearance. With the wind from the southeast, it would take a spin round to Deal to make a clean start down the English Channel. The wind later veered to the southwest, light and somewhat variable in strength. Seas were smooth as the clipper was farewelled her tug within sight of the Isle of Wight.

Shaw set course westwards close-hauled on the port tack out into the Celtic Sea. Conditions steadily worsened as the ship cleared Ushant and steered southwest across the Bay of Biscay. The wind freshened

Star of Greece

from the east as the vessel moved into the sweltering heat of the northern tropics. The immigrant ship Wild Deer outward bound to New Zealand was spoken to 300 nautical miles southwest of the Canary Islands 24 days out and asked to be reported 'All well'. The run to India was punctuated by a series of tropical storms as the Star of Greece travelled northeast towards Ceylon. Once in the intertropical zone, the trades died out as a period of dull calms and overcast skies saw the ship lolling about on a glassy sea.

Crawling slowly north, every tropical rag aloft, the Star of Greece was held up by the prevailing monsoon winds that brought with them a winter chill off the land. Despite the adverse weather Captain Shaw still managed to drag the best out of his ship and her crew. The only sad incident of note was the death of 34-year-old Donkeyman, A.G Hurle who dropped dead from heart failure on New Year's Day. His passing cast a pall over the crew's limited festivities. The ship arrived in the Saugor Roads on February 9th 1876, 95 days from Shanklin. The Star of Persia was hot on the heels of her rival. Having departed London on January 10th the crack clipper was making exceedingly rapid progress down through the Atlantic.

Loading time in Calcutta was not up to its usual rapid pace and the Star of Greece did not clear out from Ralli Brothers loading berth until March 1st. Towed downriver to Diamond Harbour those aboard the ship were surprised to find the Star of Persia anchored awaiting the flood tide, having made the run from London to Saugor Island in 77 days. With the northeast monsoon winds at her stern, the Star of Greece left Saugor on March 8th and romped on down the Bay of Bengal. The dry land breezes made for pleasant weather for much of the clipper's voyage south by west across the Indian Ocean towards Cape Agulhas, and around to the South Atlantic.

As the Star of Greece was ambling across the Atlantic the Star of Erin was making a notable run to Port Adelaide, South Australia. Having at last been repaired from her dismasting and subsequent grounding the year before the little clipper was in near new condition. The Star of Erin had left London on January 26th and passed Cape Verde on February 14th, a fine run south. The same trade winds failed to appear for the Star of Erin as she ghosted along under every scrap of canvas

Star of Greece

in an effort the make southing. Captain James Mills ran his easting down along the 39th parallel and any attempt to sail further south was frustrated by the south-easterly breezes that kept pushing the ship north against the captain's want. The run to Cape Borda was marked by a lack of wind, the same conditions experienced by those aboard the Star of Greece as she attempted to reach home. The clipper arrived at Port Adelaide on April 28th 1876, 93 days from port to port, carrying a load of general cargo, 26 passengers and a ships doctor. The Star of Erin's appearance signalled a shift in Corry's priorities as they sought to gain a toe hold in the burgeoning migrant trade to the colonies.

Sailing across the Western Ocean into light and variable northerlies the Star of Greece made slow progress to the soundings. She passed St Catherine's Point in tow on June 22nd 1876 and did not arrive off the Goodwin Sands until the morning of 24th thanks in part to the freshening of the winds causing the pilot to pause the tow until local winds moderated. Shaw and the pilot agreed that the clipper should be towed to London as they had the run of the tide. Thus the Star of Greece was entered into the lock late on the same evening. After so many months at sea, there were obvious issues with the amount of sea growth plastered to her hull. This and the fickle winds explained such a slow passage from Calcutta. The Star of Persia arrived at London on August 13th after a rather laborious run of 122 days port to port. This was by far her worst journey ever and she too had a hull in desperate need of a shave and paint. Both ships were docked for an extended period giving time for Lloyds to resurvey the ships. All of Corry's vessels as they came into port were laid up for maintenance and survey.

It was not until the last day of August 1876 that the Star of Greece was once more fully loaded and ready for sea. Captain Shaw had been away during this time overseeing the design specifications for Corry's new ship, a yet to be named clipper of some 1600 tons to be built by Harland & Wolff within the next twelve months. William Shaw had one last chance to claim the round-trip record from the Star of Persia before he left his beloved clipper for a new command.

Star of Greece

Setting forth once again from Poplar on September 2nd the Star of Greece lay at anchor in the Downs for almost 48 hours as northerly gales lashed the southeast of England. It was considered by the pilot too dangerous for the ship to depart so William Shaw and his crew were forced to wait out the blow. Rain poured down for the two days as the wind howled in from the northwest, seas were rough and confused in the Channel and many a vessel had to run into shelter to avoid being wrecked. Conditions moderated by the 4th as the front quickly passed, winds continued from the northeast but abated considerably. Seas in the Channel were now smooth with fine clear conditions expected for the next few days. The Star of Greece took her departure off Deal that morning and cracked on south-westwards towards Ushant. The wind veered sharply to the south-southwest as the ship was making her way passed the Lizard on September 6th. The glass dropped sharply as yet another blow rolled in from the west. Winds strengthened to gale force and the seas became rough to very rough as the vessel smashed her way through the Bay of Biscay.

The Star of Denmark which had left London on August 7th was gaining rapidly upon Shaw's clipper even as she was having a fast run the Cape of Good Hope. Both vessels were within a day of each other, the Star of Denmark having caught up to her competition on the voyage south. Rounding well out from Cape Agulhas both vessels were hard over under a full spread of canvas. William Shaw was having the trip of his career, his clipper fairly flying in the perfect sailing conditions. The skippers of both ships timed their runs into the Bay of Bengal to perfection as they cracked on with the quartering monsoon winds at their sterns. There was but a ship's length between the two clippers as they hove-to off of Sandheads together on November 19th, 1876. The Star of Greece had made the run from Deal to Saugor in 76 days, 81 days port to port, a fine achievement by the standard of the day. The Star of Denmark under Captain Willis had gone one better, having left Deal on September 8th she managed to catch the Star of Greece and then stay with her for most of the voyage to India. She completed the run from Deal to Saugor in 72 days, 76 days port to port. The loading times of the two ships would determine which skipper would claim not just bragging rights but the Star of Persia's record.

Star of Greece

Loading by Ralli Brother's stevedores from their barges was done at a frenetic pace and Captain's Willis and Shaw laid bets as to which vessel would make the fastest passage home. The Star of Greece was loaded and ready to leave Calcutta on December 5th, the Star of Denmark less than a day later. The race home began with tugs pulling the vessels to the Saugor Roadstead, each skipper hoped that both tug captain and pilot were worth the money being paid them. The Star of Greece shook loose her sails from Sandheads on the 9th with the Star of Denmark right behind. Captain Shaw took a southerly route pushed along by chill northerlies that swept into the Bay. The clipper crossed the middle of the Indian Ocean passing the ship Cholula bound for Calcutta from London.

With every rag aloft the Star of Greece set a cracking pace, reaching the Cape of Good Hope on January 7th, 33 days from Sandheads. Whilst not record time the ship was still flying, a bone in her teeth. Attempting to squeeze all possible speed out of her, Shaw drove her hard, keeping sail on even when conditions warranted caution. Ascension Island was passed on January 24th yet Bill Shaw did not heave to for a visit, merely passing close enough to signal his ships identity. The Star of Denmark had not yet been seen passing this way. There was still hope that the Star of Greece would arrive first.

The voyage was almost done when the ship hove to off the sleepy town of Deal on February 27th, 80 days from Saugor. The pilot cutter was signalled and a telegram was sent up to Corry's offices in London announcing Shaw's arrival. The race was still on, the Star of Denmark could appear at any time to spoil Shaw's day. A tug was quickly arranged and upon the rising of the tide, the Star of Greece was taken up the Thames and right into her berth at Poplar. All Shaw could do now was wait for the appearance of Captain Willis and the Star of Denmark.

He need not have worried about his claim to the record. The Star of Denmark was struck by calms and adverse winds for much of her voyage home. The luckless Captain Willis and his ship did not pass the Lizard until March 26th and arrived off Deal two days hence, 108 days from Saugor Island.

Star of Greece

The Wild Deer in the dry dock at Port Chalmers, Dunedin, circa 1875.
National Library of New Zealand.

Star of Greece

The Cholula, 1066 tons, built 1867 for John Willis, Liverpool, and renamed 1877 Dunnerdale.
John Oxley Library, State Library of Queensland.

Star of Greece

Captain Shaw and his crew became the toast of the town when their final voyage time was calculated and confirmed. Shaw was rewarded for his efforts and the publicity was worth countless pounds to Harland & Wolff. The record run was covered by several papers in England and Ireland. The new official 'round-trip' was on record as having taken exactly 5 months, 24 days, 21 hours.

Captain William Shaw could leave his ship with his head held high. He had a made-to-order brass game cock mounted atop the main mast signifying to all and sundry that his ship was now cock of the route. The record of the Star of Greece would never be beaten, at least not by an iron clipper. In Belfast, there were congratulations all round as Shaw arrived to a hero's welcome and was feted by both Corry's and Harland & Wolff. He was in line to take command of one of the new Star Line clippers, but both were still under construction. Thus Bill Shaw took a well-earned break from the constant stress and pressures of life as the master of a crack deep-water clipper.

Star of Greece

Cock of the Route

The new skipper of Corry's crack clipper the Star of Greece was 29 years old William Legg, a long time Corry employee. Born and raised in Carrickfergus in Northern Ireland he had joined the company's original clipper the Jane Porter as an apprentice in 1862 and had been a mate of that vessel since 1866. He was then given command of the Jane Porter in 1872. His promotion to master of the Star of Greece was seen as a vote of confidence in the young captain's abilities. As William Shaw was taking a well-earned rest Captain Legg was overseeing the loading and preparation of his ship. The vessel was contracted to take her usual 1800 tons of coal to the jute mills of Calcutta. The Star of Greece cleared out from Poplar on April 13th and was towed out to Gravesend where she lay at anchor for another for 24 hours as final preparations were complete. On the morning of April 15th 1877, the crew brought the anchor to the hawse-pipe.

Then in tow, the clipper was hauled out past Deal. She sailed down Channel having left her tug off the Isle of Wight the sea shrouded in mist and fog. As the clipper passed into open waters the winds picked up markedly from the southeast. William Legg was determined to live up to the reputation of his predecessor. He drove the clipper hard rolling down through the Atlantic and up into the Indian Ocean in near-record time. The southwest monsoon provided the Star of Greece with strong and favourable winds which dovetailed in nicely with the freshening trade winds driving the clipper rapidly on her way from Mauritius to Sandheads anchorage. The ship arrived on June 30th, 76 days from Deal. She was towed into Calcutta and was warped into her moorings on July 2nd. A somewhat bewildered William Legg was greeted with much ceremony and fuss by merchants from the local Greek community as he stepped ashore. They were perhaps

Star of Greece

expecting William Shaw to appear, however, regardless of who was in command, there were celebrations in the offing to mark the clipper's previous record-breaking voyage.

Bill Legg would have also heard the news that made him much less happy. His previous command, the clipper Jane Porter had run into some difficulty. The little ship departed London soon after Captain Legg had left, and loaded with coal had set off once again for Calcutta. Arriving off the Sandheads pilot station on June 15th, her new master Jack Legg found no pilots available. Instead, he attempted to sail into Saugor unaided. Unfortunately, Legg steered his ship too far to the east and the Jane Porter bumped heavily over a sandbank. Fearing trouble, a furious and panicky John Legg ordered both anchors let go as swinging the lead the crew found their clipper in just five fathoms of water. Those aboard considered the ship to be in great peril as the weather worsened and the ship threatened to capsize in the rough conditions. The order was given to abandon ship and the Jane Porter's boats were quickly provisioned and taken out onto the davits. Collecting their personal items, the crew took to the boats, the captain taking charts, navigation instruments and his and the ship's papers. Last over the side, he joined the luckless sailors as they rowed away from the stricken vessel. The crew spent several days rowing up the Matla River to Port Canning from where they travelled overland to Calcutta.

The town of Canning had existed as a deep-water river port for just a few years. The idea for a port on the River Matla had been mooted for some time and was pushed through by Lord Canning, after whom the town was named. In 1862, the Eastern Bengal Railways extended the line from Calcutta to the river in preparation and soon a series of monumental stone and wooden municipal buildings were established on the site of the former fishing village. The port town lasted until the 1867 cyclone all but wiped it off of the map. The river had surged and overflowed its banks sweeping much of the town and its newly built port facilities away. There were efforts to rebuild the port that lasted until 1870 when the Port Canning Trust went bankrupt.

Star of Greece

An example of the carved wooden panel presented to Captain Legg by the Greek jute merchants. These were then fixed within the cabins of the clipper.
Aldinga Library Collection.

National Flag of Greece, circa 1860s.
(White cross on a light blue background, blue and white checked shield with a golden and ruby-encrusted crown atop.)

Star of Greece

When they reached Calcutta on the afternoon of July 15th Jack Legg immediately notified Corry's agents who despatched a steam tug to investigate the current state and position of the Jane Porter. The tug's crew found that the clipper was still safely at anchor in the mouth of the Matla River. With not enough crew to raise the ship's anchors, both cables were slipped to tow the Jane Porter across to Diamond Harbour. The tug hauled the ship around the myriad sandbars and shoals, up through the proper channel of the Hooghly and into Diamond Harbour where it was discovered that she had suffered almost no damage from her time abandoned in the Matla mouth. The master of the Jane Porter would have some very serious questions to answer from both Corry's and the Board of Trade when he arrived back in London. It can only be wondered at when trying to guess William Legg's reaction when he arrived in Calcutta. He had other more entertaining concerns at the time with festivities, honouring the Star of Greece's achievements, taking place for much of the time she was in port.

Ashore in Calcutta Captain Legg and his officers were invited to a series of almost end dinners and soirées hosted by the merchants from Calcutta's Greek community. They celebrated the wonderful records set by the clipper and presented William Legg with a silken flag representing the Kingdom of Greece and wooden carvings that were to be mounted inside the saloon. These gifts and Bill Shaw's brass gamecock atop the main truck announced to the world that the Star of Greece had, at last, claimed the record from the Star of Persia. The ship was now 'Cock of the Route' between Calcutta and London. The flag was to be flown whenever the Star of Greece was in the Hooghly and the gamecock was mounted atop the mainmast, a trophy for all to see. Chief amongst these to support the success of the clipper were Ralli Brothers, jute mill owners, exporters and most importantly Corry's main supplier in Bengal.

The Greek diaspora had been in Calcutta since the early 18th century. The community grew rapidly when many families migrated there from the rich Thracian cities of Adrianoupolis and Philippoupolis during the Turko-Russian war of 1774. More migrants arrived with British colonists from the Ionian Islands as well as from the Greek cities in Cappadocia and the Aegean.

Stephen A Ralli. Circa 1870.

Star of Greece

By the late 19th century the community was centred on 120 families mostly living in central Calcutta. Many Greek families engaged in trade and became very prosperous. During the latter years of the century, there were several trading companies operated by very well established Greek families. These included Ralli and Mavrojani, Argenti Sechiari, Agelesto Sagrandi, Schlizzi and Co, Petrocochino Bros, Tamv-aco and Co, Georgiardi and Co, N Valetta and Co, Giffo and Co, Pallachi and Co, Vlasto and Co and Nicachi and Co. Preeminent amongst all these houses was that operated by the Ralli brothers.

The Ralli's opened their first Indian office in 1851 at 15 Lal Bazaar, Calcutta. They began by buying jute from press owners and then exporting it on sailing ships bound for Europe. The firm was originally founded in London in 1826 by Pandius Ralli whose family hailed from the Aegean Island of Chios. The firm quickly expanded, owning its own ships it exported many thousands of tons of cereals, foodstuffs, spices, cloth and other commodities, from India and the Levant to all over Europe. From 1866, under the management of the brothers Stephen and John Ralli, the company of Ralli Brothers Calcutta Incorporated, exported enormous quantities of jute to most countries in Europe, especially Britain, Germany, France, Belgium, Holland and even to Russia when the jute crisis there began to bite. Ralli Brothers business grew to the point where they needed to guarantee regular supplies of jute to meet the demands of their European customers. Thus, they built factories for cleaning and finishing jute and the pressing of bales for export. Their industrial and commercial ventures earned the firm millions of rupees and made the Rallis one of the richest families in Calcutta.

If goods were wrapped or bagged, the Rallis were involved. Even though jute was at the heart of their business empire, Ralli Brothers offices in Calcutta were also heavily involved in the manufacture and export of shellac, turmeric, ginger, rice, saltpetre, borax, cotton thread and manchester. Riding on the boom in the jute trade Corry & Co' and Ralli Brothers carved out a sizable niche for themselves making large sums of money for their respective shareholders. Much of the success of their joint venture was down to the quality of the ships and crews used to tranship the jute to Europe and North America.

Star of Greece

With her new accoutrements, the Star of Greece was warped out from Ralli's loading berth and floated on down the Hooghly. William Legg farewelled the local pilot off of Sandheads on August 6th 1877, and with the monsoon winds holding steady from the southwest, set course south by southeast skirting the Burmese side of the Bay of Bengal. Tracking along the tea clipper route the Star of Greece picked up a favourable slant and rocketed southwest towards the Cape of Good Hope and home.

Even as Captain Legg was bringing his clipper around the southern tip of Africa Harland and Wolff were handing over the latest addition to Corry's fleet, a newly completed ship, the 1644-ton Star of Italy. The Star Liners were normally consigned to sail from Belfast to Liverpool to take on their first cargo. In this instance, a steamer brought 2000 tons of salt to Belfast which was then loaded into the clipper's three holds for shipment to Calcutta. With William Shaw in command, Corry's could certain that the ship would perform up to expectations. The start of her career in the East India trade was less than stellar. Like the Star of Albion that had been in a collision because of violent gales a month before, the Star of Italy came to grief when her anchor cables parted and she collided with the barque Ellen. Both vessels suffered extensive damage to bulwarks and rigging and had to be slipped for inspection of their hulls and repairs. Such was the ferocity of the gales that roofs were torn off and trees uprooted all across Belfast. The winter proper had not yet begun and already Corry's shipping repair bills were eye-watering.

William Legg and the Star of Greece were having no such troubles. The clipper was making fine time rolling home with fair winds in the South Atlantic. Ascension Island was passed on October 22nd, 77 days out, less than a day behind the brand new clipper Star of Bengal, which departed from Saugor on July 27th. The two clippers old and new kept pace with each other as they raced up the north Atlantic into the teeth of a northerly gale. The voyage from the Western Isles to the Scillies was punctuated by force ten winds and confused seas.

Star of Greece

The Star of Italy, 1644 tons.
State Library of South Australia.

Star of Greece

By the time the Star Liners sailed into the western approaches conditions had abated markedly so that now light south-easterly winds and smooth seas greeted the ships and their exhausted crews. The Star of Bengal slipped into the English Channel less than two days ahead of the Star of Greece and hove to off of Deal on November 3rd, 99 days out. William Legg was happy to pick up the pilot off of the Isle of Wight having passed Prawle Point on the morning of November 5th. The Star of Greece hove to off Deal the next morning, 92 days from Saugor, as a side-wheeler came out to haul the clipper into her Thames anchorage. The Star of Bengal had managed to pull away from the Star of Greece after Ascension Island but Captain Legg's seamanship once again showed the true qualities of his ten-year-old ship. The vessel eventually entered the East India Docks on the evening of November 7th, there she was warped into her berth alongside the other Corry clippers; the Star of Bengal, Star of Erin and William Legg's old ship the Jane Porter.

Loading continued apace through November and well into December as Harland and Wolff completed yet another ship for Corry's fleet. The 1600-ton Star of France was the twelfth and last clipper ordered from Queens Island. She would not be ready for service for some weeks as the fitters and riggers prepared the ship for sea. Much of the rest of Corry's fleet was berthed in the East India Dock, awaiting charters before moving over to the Poplar coal chutes to finish loading for the run to India. The Star of Bengal was the first clipper away departing London on December 5th, the Star of Greece languished in the dock for all of January as Corry's worked out new contracts with jute suppliers and mill owners in Dundee.

There was a growing trend for jute millers and exporters in Calcutta to send their products to Europe by steamer. It was now more of a battle for Corry's to get the freights they were used to from London importers and thus they waited until new markets and shipping contracts could be negotiated. Rallis were opening up the North American market and there were brand new routes to exploit by Corry's and their business partners in India. The Stars of Denmark and Albion along with the Jane Porter had already departed when loading was finished for the Star of Greece. The clipper was not cleared out from London until February 23rd 1878, having lain idle

Star of Greece

since the start of December the year before. The ship dropped anchor in the Downs roadstead on the evening of February 25th not taking her departure until the next day.

Dungeness light station was signalled at noon. The light and variable south to southwest winds meant that the Star of Greece did not drop her tow until well past the Isle of Wight when the pilot took his leave from the clipper as well. Once out into the North Atlantic the Star of Greece rolled rapidly south as an intense low pushed in from the southwest bringing with it howling north-westerly winds and huge seas. The run to Calcutta was routine with Cape Verde being passed on May 1st. The ship continued heading south encountering the usual storms off of Africa's southern coast. With the monsoon winds in a period of flux, light airs and offshore winds made for slow sailing as the clipper coasted into the approaches of Sandheads, on May 25th, 89 days from Dungeness.

The anchorage at Garden Reach was filled with watercraft of all shapes and sizes. The bottom had dropped out of the jute market this year as poor-quality jute and an oversupply meant that many a vessel was laid up for weeks at a time until an economical freight rate could be negotiated. There were over 100 000 tons of shipping in the Hooghly waiting to load. Corry's clippers arrived at their loading berths offering to ship jute at 37s 6d per ton. The Rallis and their compatriots were having none of this and thus Captain Legg was forced to accept the rate of 30 shillings. Sales in London were almost non-existent with small lots of low-quality jute being offered. Demand from mills in London had all but disappeared as Dundee took over the lion's share of the milling of all qualities of jute. Corry's agents organised for the Star of Greece to load 8571 bales of poorer quality jute. The Star of Greece was laid up in the Hooghly alongside the Jane Porter until July 9th 1878 when loading was completed. Then outward bound from the Hooghly William Legg farewelled the tug and the pilot on July 16th sailing for Dundee.

Star of Greece

**Postcard of farmers bringing raw jute to the mill for processing.
Circa 1870's.**

Star of Greece

Ships in Dundee Harbour, circa 1875.
Photographer-Alexander Wilson, jute mill supervisor.
Dundee Central Library.

Coal Staithes South Shields, c.1890.
Newcastle Libraries.

Star of Greece

Captain Legg set a course southwest, taking the Star of Greece along the Coromandel Coast, utilising the land breezes to drive his ship south. With fresh and warm quartering winds the Star of Greece cleared Ceylon's east coast in a little over a week. From there through the expected calms the clipper raced away south to the trade winds and onto Cape Agulhas. It was whilst the ship was ghosting along through the tropics that 20-year-old Michael Lane collapsed with sunstroke in the stifling heat and humidity to the east of Mauritius. The poor unfortunate succumbed on August 7th and was buried at sea.

High seas and dirty weather prevailed as the Star of Greece romped on up the Channel under topsails and headsails. Deal was passed on October 28th as the vessel battled a bitterly cold northwest wind. Patches of fog and rough to very rough seas greeted the ship as she passed into the North Sea giving the crew no chance to dry out. Once into open waters, the winds picked up and seas became more violent and confused as a low-pressure system moved in over the North Sea. The cold front soon passed and yet the Star of Greece had to contend against a dead muzzler for much of her way to Dundee. The light and variable northwest winds forced William Legg to sail well out to sea before hooking around to westward and into the mouth of the River Tay. The ship reached the Taymouth roads on November 9th to await the arrival of the Tayport pilot and a side-wheeler to bring her safely into Dundee harbour.

The Star of Greece was warped into Dundee's Victoria Dock to discharge her cargo which was destined for the mills of the rapidly growing city. It had taken almost two weeks to reach Dundee from Deal due to the adverse weather. Her run from Saugor to the Tay Mouth took 116 days, 121 days port to port. Aside from her cargo of jute Captain Legg brought home for his personal use three cases of Indian tea, and several dozen boxes of cigars. The Star of Greece was held up in Dundee Harbour for almost a month whilst Corry's agents D. Bruce & Co' organised a return cargo. Legg received orders on December 3rd to sail for North Shields to take on coal for Calcutta.

The Star of Greece departed Dundee on December 5th in ballast and was towed down the coast to Shields harbour on the River Tyne. Both vessels arrived a day later on December 6th, and the ship was taken to

Star of Greece

the inward mooring buoys so that the crew could make preparations for coal to be taken on board. The lower masts were cockbilled to allow the ship to be taken in beneath the coal staithes. These were short piers that projected out over the River Tyne and allowed coal wagons to run on rails to the end. The Star of Greece was moored alongside and the coal from the wagons was emptied down chutes into her three holds. Loading was completed by December 17th and the soot-covered vessel was cleared out the next morning. With final sea stores aboard the ship was towed out to the anchorage on the evening of the 18th. Captain Legg gave the order to heave up the anchor early the following morning and soon the Star of Greece was underway southward under tow, bound for Deal and onwards to Calcutta. As the clipper was heading south, a great rival, the Star of Bengal was readying to set sail from London. Both vessels passed through the Downs within an hour of each other and crossed paths more than once as they raced onto the Western Isles.

Whilst the Star of Greece and the Star of Bengal were charging south towards the equator economic forces were at work that would have a lasting impact on the future viability of using sailing ships to transport jute from Calcutta. The main economic drivers for this shift were the jute mill owners in Dundee. Due in part to the drought that had gripped much of India and the plummeting freight prices mill and steamship owners had banded together to open up a new jute growing district centred around the deep-water port of Chittagong. It was a well-established fact that Calcutta was the most expensive port to ship anything from on the east coast of India. Pilotage, towage and port charges were all exceedingly high compared to the rest of the subcontinent adding as much as 15 shillings per ton for jute shipped to Britain.

There were many advantages shipping jute from Chittagong. The port was no further away from the jute growing districts, ships avoided the dangers of the Hooghly, there were not the expenses of Calcutta port charges and towage fees, pilotage, lighterage and harbour dues were trifling when compared to Calcutta. Chittagong itself possessed a safe, deep water anchorage and there were many bye-waters of the Ganges down which raw jute could be economically transported from the interior by growers. There was a push for mill owners to build

Star of Greece

new jute presses to allow direct transhipment to steamers for the conveyance of jute up through the Suez Canal, bypassing the lengthy sea passages currently experienced by the clippers sailing from Calcutta. It was a simple matter of supply and demand; a steamer could take ten to twenty thousand bales of jute direct to London or Dundee in around forty days. The fastest clippers which handled less than half these amounts travelled via the Cape of Good Hope in around eighty to one hundred and twenty days.

Two Corry clippers raced on south searching for the fitful breezes of the northeast trades. Captain Legg followed Maury's directions sailing towards the northeast coast of Brazil to make the shortest run across the doldrums, passing Cabo Sao Roque on January 26th 1879, 38 days from the Downs. They stayed in sight of one another for much of their voyage south around the Cape of Good Hope, making their easting turn along the 38th to 40th parallels before swinging their courses north across the centre of the India Ocean. It was a wise choice as the clear signs of a violent tropical storm appeared well off to the west, a cyclone that would smash into Mauritius in the coming days. It was a master class in seamanship by Captains Legg and Smyth that saw the Star of Greece arrive off the Sandheads pilot station just a few hours before the Star of Bengal, both vessels 108 days from the South Downs.

They had been anchored in the Hooghly for eight days when tragedy struck. The ship's steward, 22-year-old Robert Petrie fell overboard into the murky polluted waters of the river. Arms full of supplies for the captain's cabin the young man slipped from a narrow gangplank connecting vessels moored side by side. Unable to swim Petrie became trapped in the muddy waters between the two vessels. Despite the best efforts of those aboard the unfortunate young man was drowned before he could be rescued. Despite this setback, loading was completed promptly and the clipper was moved out from Calcutta on May 25th. The Star of Bengal had been loaded by May 20th and had sailed from Saugor on the 24th. Bill Legg felt a certain drive to overhaul the Bengal and his friend Captain Smyth. There was little contest between the two vessels with the Star of Bengal getting the better of the winds for much of the journey home. Captain Smyth's ship passed the Lizard on August 26th, after 94 days at sea.

Star of Greece

The Star of Greece was well behind having struck light airs as she entered the northeast trades. Her run into the western approaches was faster as she thumped along ahead of a quartering westerly gale. Howling winds and driving rains accompanied the Star of Greece as she sailed into the Channel. The appalling weather conditions forced Captain Legg to sail close in under the Lizard light station, on September 9th, signalling his ships safe arrival, 105 days from Diamond Harbour. The Star of Greece sailed onwards to the Goodwins where she dropped anchor on the morning of the 10th to await her tow. It was not a trip which provided much in the way of a profit for Corry's. Freight rates had dropped to below 30 shillings per ton and other cargos and destinations were being searched out by the company's shipping agents.

Having been laid up for a time as jute rates plummeted to an all-time low Corry's agents managed to arrange for the Star of Greece to sail out to New York. The charter stated she was to take on a consignment of wheat destined for the flour mills of Liverpool. She was warped into her East India berth to take on a load of chalk. Considered as naught but ballast by most ship owners, the chalk was of such fine quality that E.L Lawrence & Co', silversmiths and pen makers of New York were more than happy receive the chalk for use as an abrasive and silverware storage medium for their factory and warehouse. William Legg had his vessel hauled down the Thames on September 28th, 1879 looking like a ghost ship covered as she was in a fine layer of chalk dust. Deal was passed on the 30th as the Star of Greece set sail into a dead muzzler as she dropped her tow off Shanklin, Isle of Wight.

Cracking on down the Channel Bill Legg was faced with little choice, prevailing westerlies meant that he was forced to take the Star of Greece on a zigzag trip across the Atlantic. A voyage that took the average steamer about ten to 12 days became an arduous struggle of 31 days to Sandy Hook pilot station. Hove-to in a westerly gale the clipper bobbed about unhappily as she waited for the snappy little pilot schooner to shoot her way across the bar. Soon the doughty little craft arrived to drop off one of the many expert local mariners who would guide the Star of Greece across the sand bar and into the

Star of Greece

roadstead of Lower New York Bay. From there a steamer towed the rather crank clipper into her East River berth to discharge her by now an almost solid lump of chalk. Once unshipped the fine white powder was to be taken Lawrence & Co's warehouse.

The Star of Greece was warped into her discharge berth on November 3rd 1879 and was in port for a little over three weeks. She cleared out from the East River on November 26th, her hold filled with more than 1500 tons of bagged wheat, just a day behind the Star of Persia. Bill Legg had his ship towed out to the Lower East River and from there with the pilot discharged she was up and away from Sandy Hook on December 1st. Flying before the westerly gales she made a rather wet voyage back across the North Atlantic. The southern tip of Ireland was sighted on the 17th and the Star of Greece rolled into St. Georges Channel making the Mersey roadstead on December 19th after a rapid-fire run of 18 days. The ship was brought up the Mersey to the Liverpool docks to discharge the following morning. Christmas was spent dockside with Legg having to deal with the Liverpool boarding house crimps to replace those members of the crew who had run or taken their leave from the ship.

Thankfully for Captain Legg, the Star of Greece was in port just long enough to take on coal for Calcutta. She cleared out from Liverpool on January 13th to be towed out to the roads taking her took her leave of the smog-filled harbour the following day. The Sand Heads pilot's station was reached on April 18th, 95 days from Liverpool. The freight margins had improved only marginally on the previous year even though the rebellion in Burma had threatened the development of Chittagong. It was whilst she was laid up at her moorings that Captain James Mahood and his clipper the Star of Persia arrived from London. The two ships were joined by several other Corry clippers all awaiting a rise in the rates which would make their runs home profitable for all concerned. Captain Legg got his ship under tow on June 5th for a run down to Saugor which was reached on June 7th. The Star of Persia followed the Star of Greece down the Hooghly on June 28th and said farewell to her pilot on the 30th.

Star of Greece

New York dock--Stevedores unloading a ship, circa 1877.
Harper's Weekly, 1877 July 14), p. 540. Library of Congress
Washington, D.C.

Star of Greece

After battling the westerly gales off the Cape of Good Hope the Star of Greece experienced a fine weather passage to the line, with Cape Verde being sighted on September 12th, 95 days out. Becalmed for a time the Star of Greece was being closed upon by her rival. The trade winds proved rather strong blowing from the east and the north. Bill Legg pushed his vessel out to the middle of the North Atlantic in search of the prevailing westerlies that would take his ship home. He took the clipper to 50° north, passing the Compagnie Générale Transatlantique 3420-ton auxiliary steamer St Laurent before the winds swung in his favour. Hooking around to starboard the Star of Greece raced on home picking up the Deal light on October 5th, 118 days from Saugor Island. This was her lengthiest run ever and the duration of the voyage had brought the ship's owners and agents more than a little concern for her safety. James Mahood aboard the Star of Persia made a much tidier run home, making Dover on October 7th 1880, 99 days pilot to pilot.

Once home the Star of Greece was directed to the West India Docks on the Isle of Dogs to discharge. It was a first for Captain Legg and something of a novelty to unload there. He felt that there was a change in the wind, that whilst he had been away Corry's had shifted the focus of their business. With the current depression in the jute market, Corry & Co decided there was much more money to be made in the burgeoning trade to the colonies. Wool and grain were becoming the staple cargoes for iron-hulled clippers and the Star of Greece was contracted to Devitt & Moore's Australian Line. The docks covered nearly 300 acres between Limehouse and Blackwall on the north side of the Isle of Dogs. They were large enough to berth more than 450 vessels. The shore was lined with warehouses in long rows along the banks of the Thames several of which were consigned to the cargoes of Devitt & Moore. The load going into the Star of Greece was in large part made up of iron wares essential for the growth of the colony of South Australia.

The West India Docks.
George Birch, London, 1895.

Star of Greece

The truly varied cargo also was made up of haberdashery, clothing and millinery, glassware and window panes, agricultural implements, wire and wire rope, leather goods, bags of seed, toys, pianos, machinery, carpets, china and earthenware, barrels, bottles and casks of beer and spirits, pepper, spices and cocoa, vinegar, cigars, vegetable oils, peas, tinned fish, soda, saltpetre, coal and many other lines of general hardware. The eclectic mix of goods required special preparations made in the lower holds and shifting boards, partitions and canvas coverings applied to the 'tween deck spaces. Loading was completed by December 17th and the Star of Greece was cleared and towed down to Gravesend on December 20th. With her pilot aboard, she was brought out to the Downs on the morning tide and departed Deal the next day bound for Port Adelaide.

A report in the South Australian newspaper taken from William Leggs' sea-log and his observations described her voyage thus;

"Star left London on December 20 and had very fine weather down Channel, and then fell in with a brush of fresh breezes, but nothing worth noting. After six days she was quite clear of the land, but made no very good runs, as she reached away to the southward, and the north-east trades proved so miserably scant that she was thirty-seven days to the Equator, which was crossed in 25° 20' west. The southerly trades proved better, and on February 3 she sighted the Island of Trinidad and passed within ten miles of it. On crossing the prime meridian she headed well to the southward till reaching 40° and on March 3 had the Island of St Paul bearing east distant seventy miles. Thence to Cape Leeuwin had tolerable weather, but then the wind came out from eastward and gave her a dead beat up to Borda. On March 20 she fell in with the outward-bound steamer Aconcagua, and from her obtained the bearings and distance of Cape Borda. On the 22nd she passed the barque Saraca hence, and in due course reached the outer harbour, when the pilot boarded and having arranged for the attendance of a couple of tugs, she was towed up on the evening's tide." **The South Australian Advertiser Thursday 24 March 1881.**

On her outward voyage, the Star of Greece passed and spoke to several vessels. On December 29th well west of the Bay of Biscay, she passed the Nereus and asked to be reported 'All well.' On January

Star of Greece

13th at 21° N, 31° W she spoke to the steamer Copernicus out of Buenos Ayres and the clipper Earl of Granville bound for Brisbane and again reported 'All well' and 56 days out from London approaching the line of the Cape of Good Hope, she spoke to and passed the clipper Bowden from London also bound for Port Adelaide. The Star of Greece was moored at Outer Harbour on March 23rd, 92 days from Deal.

Much of her cargo was consigned to Harold Brothers, Devitt & Moore's agents in Port Adelaide and the ship was warped into Quay 1, to discharge. William Legg and his apprentices had plenty of time for a run ashore as they awaited the arrival of ketches bringing in their loads of wool from South Australia's outports. The diverse nature of the cargo meant for a slow and careful discharge as consignees fronted the wharf over the coming weeks to retrieve their merchandise from either the ship or Harold Brother's warehouse. Once empty the clipper was warped out and towed back into the outer harbour where a series of cargo lighters and ketches waited to tranship their loads of wool and grain into her holds. Thanks to the lumpers of the South Australian Stevedoring Company loading was completed by May 10th 1881 and the Star of Greece was cleared out.

The various crimps of the Port Adelaide boarding houses were the key recruiters, managing to sign on a worthy group of sailors for the run home, Bill Legg was forced to pay their exorbitant fees to guarantee the pick of sailors between berths. Chief amongst those who profited greatly from the practice of crimping were Holman & Wallen, boarding housemasters of Port Adelaide. They ran a thriving business accommodating sailor's fresh from the ships that docked in the port and were instrumental in organising the supply of men for outbound vessels. One of the sailors who had jumped ship in Adelaide was a young man named Andrew Peterson. The unfortunate sailor had left the ship and had been quickly snapped up by the local crimp and shipped out on board the 100-ton schooner Experiment.

Star of Greece

Port Adelaide looking from the Birkenhead side of the Port River.
State Library of South Australia.

Star of Greece

The luckless seaman was drowned along with the vessels master George Hunt when the Experiment was hit by a violent squall, heeled over and sank off Althorpe Island in less than two minutes on May 11th, at 2 o'clock in the morning. Two survivors managed to take to the schooner's small boat before she sank and were picked up the next day by the schooner Grace Darling. The fortunate men were then transferred to the steamer Hawthorn for passage back to Port Adelaide.

A heavily laden Star of Greece was cleared out with 9656 bags of wheat, 702 bags of flour, 6676 bags of bark (for tanning), 203 bales of wool, 146 casks of tallow and 39 bales of skins aboard. William Legg had bought 5 casks of local wine for himself and these were entered on the ships manifest for export. Harold Brothers had managed to secure the shipping rate of a flat 45 shillings per ton. Whilst not great it was still a much better haulage rate than the 30 shillings per ton of jute being offered in Calcutta. With such a cargo she rode well down on her marks. In the company of a pair of tugs the Star of Greece was towed out to the Semaphore roadstead on completion of loading and with a full complement of sailors and idlers aboard set sail down St Vincent's Gulf on May 13th, London bound. The clipper sailed towards Kangaroo Island with a quartering wind from the north-northwest. All indications were that a blow was on the way from the southwest convincing Bill Legg to take the longer, rougher, and quicker route around Cape Horn rather than try to beat westward to the Cape of Good Hope.

Stormy, gale force winds blew almost to Cape Horn, Drake Passage being reached just a month after departing from Port Adelaide. Plunging to more than 50° south the Star of Greece sailed on through the icy conditions battling more than one Cape Horn Snorter with their huge spume topped greybeards. The bitter and icy conditions did not begin to abate until the ship was well east of the Falkland Islands. The run up through the Atlantic was equally as fast with Cape Verde being reached after just 74 days at sea. It looked to those aboard the Star of Greece that they may make it home in less than a month. It was not to be.

Star of Greece

Port Adelaide: South Australian Company's wharf, on the left is Levi's Wharf, circa 1867.
Port Adelaide Collection, State Library of South Australia.

Star of Greece

Within a few hard days sailing of England, the winds began to blow steadily from the southwest. Captain Legg ordered the crew aloft to set all plain sail as headway was finally made eastwards. Sailing in past the Scillies the wind veered round to the west-northwest becoming light and unsettled as periods of rain and the occasional thunderstorm drifted in over the southwest of England. The Star of Greece cruised on by the Lizard on the morning of August 26th 1881, 105 days from Semaphore. With a freshening south-westerly at her stern, the vessel romped home all the way to Deal anchoring in the Downs that same evening. The Star of Persia still commanded by James Mahood sailed past the Wolf Rock light later that same morning and anchored in the Downs not far from her sister. Both vessels took up their tows and were hauled down to Gravesend to await an opening at the docks in London. The Star of Greece was destined for the Isle of Dogs whilst James Mahood's clipper was to be sent home to Poplar. As the two ships lay at anchor in the Thames the Star of Russia under Captain Smyth and the Star of Bengal with John Simpson in command passed the Lizard on their way to London from Calcutta. All four clippers were entered into their respective London berths to discharge by August 30th after lengthy voyages all round.

Such was the success of the run to Adelaide that Corry's again contracted the Star of Greece to sail out to Australia. This time she was Sydney bound loaded with general cargo for the colony. Soon William Legg's ship was joined by the Star of Persia in the South West India Docks, bound for Adelaide. Loading was completed for the Star of Greece by October 9th and those aboard her said farewell to London the next morning. With the wind blowing in from the west the Star of Greece set sail from the Downs at 7:00 am on October 13th. The clipper sailed straight into the teeth of a ferocious westerly gale accompanied by high seas and driving rain that at times reduced visibility to just a few hundred yards. Dozens of vessels of all sizes were caught out by the unexpected intensity of the storm.

Less than 24 hours out, having lost five sails blown clean out of their bolt ropes, Captain Legg decided to come about and run back for the shelter of Dungeness. Even as the Star of Greece rounded the cape to seek shelter in the East Roads, many other vessels were coming to

Star of Greece

grief. The barque Harmony was driven ashore at Dunkirk her cables having parted, the brig Susan of Montrose was wrecked on Holy Island, her crew of seven saved by local fishermen, whilst the schooner Midas foundered near New Brighton her captain drowned. Back in London, the Star of Persia with no stiffening aboard, other than her 150 tons of kentledge, parted from her moorings and capsized in the docks coming to rest atop several loading barges moored alongside.

The hurricane blew throughout the day, causing those aboard the now anchored Star of Greece more than a modicum of grief. With all sails taken in, and bower anchor set down, the ship lay in the East Roads with dozens of other vessels, her head into the wind. At the height of the storm there came a shriek of panic from those on watch as the anchor began to drag. Legg ordered more cable to be paid out to take the pressure off the single anchor. The first mate's watch was working frantically upon the fo'c'sle head preparing to let the second bower anchor go. The ship gave a sudden lurch as the cable parted leaving the Star of Greece to drift helplessly. The mate swung an iron mallet releasing the second anchor and many fathoms of cable. Thankfully for all aboard the second hook bit deep into the mud and the Star of Greece managed to ride out the blow with little more damage. They were amongst the lucky ones, for all across the south of England hundreds of houses were unroofed or damaged, there was local flash flooding, dozens of vessels were either driven ashore or foundered, and hundreds of people were injured or killed.

Captain Legg was faced with a sizable repair bill as the weather subsided and seas calmed. Besides the lost anchor, the Star of Greece was short 68 fathoms of anchor chain and 5 sails amongst other items washed overboard. A tow was arranged to take the ship back to the Downs. Once safely moored a new anchor, 70 fathoms of cable, sails, rigging wire and other damaged or lost items were sent out. The crew worked feverishly for the next three days and such was the diligence and skill of their labours, the Star of Greece was able to get back underway on the morning of October 18th. The Star of Persia was eventually righted by crane and with extra ballast aboard was taken into dry dock for inspection and repairs.

Star of Greece

An article appearing in the Sydney Morning Herald described the Star of Greece's voyage to Sydney much of it in William Legg's own words,

"The Start Point was passed on the 19th, and very bad weather was experienced outside the 'Chops of the Channel'. To quote Captain Legg's words, 'We had all sorts of weather, gales with thunder and lightning, and the barometer down to 28.0. After the gales had passed away fine weather ensued, which lasted seven days with fair winds. The N.E. trades were got in 18° N, but were very unsteady and light, but they were carried to the point of the S.E. trades, which were caught in 4° N. The Equator was crossed on the 20th November in long' 20.20° W., and the trades were carried down to 25° S., and then fine weather set in lasting to the Cape with moderate winds. The ship passed the meridian of the Cape on December 13, in lat. 40.53° and the easting was run down between 40° and 45°, with winds mostly from W and NW to SW. Tasmania was rounded on January 9 and gales were encountered on the 11th and 12th instant. On the 13th there were strong westerly gales and subsequently strong N.E. breezes. Favourable breezes set in about 3 a.m. on the 16th, and brought the vessel to port on the evening of that day."
The Sydney Morning Herald Tuesday 17 January 1882.

The Star of Greece was hauled into her berth to discharge her mixed cargo valued at £46700, containing amongst other things china and earthenware, toys, Manilla bags, furniture, picture frames, musical instruments, cutlery, firearms, sporting goods, stationery, perfumes, ironmongery, saddler, leather goods, tools and hardware, fireworks, medicines and medical equipment, soap and ladies toiletries, tobacco products and a variety of other goods.

Once her cargo was discharged the clipper took on ballast for stiffening and was cleared into the coal loading wharf on February 23rd. A tug was engaged on February 25th and the Star of Greece was towed out through Sydney Heads bound for Calcutta with 1000 tons of coal aboard. Captain Legg made his course south by southwest against a gentle south-easterly breeze on smooth seas. He was forced to tack and wear ship when beating his way down the east coast heading for Bass Strait.

Star of Greece

Sydney Harbour, circa 1882.
State Library of New South Wales.

Star of Greece

The light southerlies continued as the clipper cleared Gabo Island when the wind backed around to the east allowing William Legg to take his ship through Bass Strait and onto Cape Otway. Ghosting along ahead of gentle easterly breezes the Star of Greece made good time as she cleared King Island sailing once again into open waters. As the high-pressure system passed east the wind came around to the north-northwest ahead of an approaching southerly change. There was little to the change as the wind came around from the south the winds light and unsteady before again backing around from the west-northwest. The Star of Greece at times lay becalmed as the winds failed altogether leaving her bobbing about in the Great Australian Bite. With yet another high overhead the Star of Greece eventually made Cape Leeuwin on a zephyr of a north-easterly breeze which allowed William Legg to sail slowly north by west in search of the south-easterly trade winds. The trades were picked up at 25° north and the Star of Greece cruised north by west encountering a series of intense tropical storms as she passed west of Anjer, Java. The northeast monsoon was blowing steadily as the ship entered the southern reaches of the Bay of Bengal. The Saugor Roads were reached on April 29th, 63 days out.

The oppressive heat of the Bay of Bengal gave way to the cooling winds that caressed Calcutta at this time of year. Freight rates had improved only marginally for jute on the previous twelve months yet they were high enough to make the voyage home worthwhile so long as William Legg could keep the Star of Greece's incidental overheads to a minimum. The pilotage, towage and mooring fees were unavoidable and ate up almost half of the cost of shipping from Calcutta. These costs had already been factored in by Corry's agents. What could be avoided were the unnecessary costs attached to driving the ship hard on the run home. Repairs were costly and profit margins tight all across the sailing ship world. Loading was not completed for almost five weeks. This proved somewhat frustrating for Captain Legg as he was under constant pressure to get away as quickly as possible even at the height of the Indian cyclone season.

The Star of Greece left Calcutta on June 12th The breezes proved extremely light and uncertain for much of the voyage south to the Cape. Arriving off the southern tip of Africa in the middle of winter

Star of Greece

Captain Legg and his crew were faced with a series of hellish westerly gales. The clipper was held up for quite sometime before Bill Legg found a favourable slant that allowed him to take his ship north by west into the South Atlantic. St Helena was passed on September 14th, 92 days from Sandheads. With her bottom completely fouled with weeds and baffling headwinds for much of the trip, this was turning into the slowest voyage of the clipper's otherwise stellar career. The winds for her run north proved much more favourable and with continuing fine weather the ship finally made the English Channel on October 13th, anchoring in the Downs three days later, 124 days from Sandheads pilot station.

The jute brought to London by the Star of Greece fetched a higher premium than that shipped later on. Her jute came in dry and thus she was able to load a greater price per ton than the jute that had come in on her sister ship the Star of Persia. Her bales of jute had been stowed aboard at the height of the wet season and came to London sodden and damaged. Freights had none the less risen considerably and now Corry's ships were bringing in the current jute crop at 55 shillings per ton. The Star of Albion meanwhile was laid-up on the slip with major stern damage in Calcutta having run aground after colliding with a tug in the Hooghly back in September and was not expected until the new year.

After her cargo was discharged William Legg put the Star of Greece into dry dock for a clean and paint, her bottom so fouled it looked more like a farmer's meadow. Back on the Calcutta run the Star of Greece was warped into her East India berth to take on coal for Bengal. Jute rates had dropped again, the Star of France loading in Calcutta at just 43 shillings per ton. The Star of Persia arrived from Calcutta later in the month and was placed on the berth for Sydney, taking on board a general cargo in the West India Dock.

The Star of Greece set sail once more on November 23rd from London. Freshening headwinds from the west, southwest made for an onerous beat down the Channel. Prawle Point was not reached until November 30th at which time the tug and pilot were farewelled as the ship experienced rough to very rough seas to the Chops of the Channel. Once into open waters, the westerly gales strengthened in

Star of Greece

force bringing with them snow showers, hail and rain. Captain Legg had made his run up through the middle of the Bay to allow for the strengthening northeast monsoon. The pilot brig was met on March 15th 1883, 110 days from Deal.

Upon arrival in Calcutta William Legg came to learn that jute freights had dropped to 35 shillings per ton, there was little incentive for a quick and hazardous trip back to London. The stay in Calcutta was surprisingly brief, the Star of Greece departing the Garden Reach on April 20th. The north-easterlies blew cool and fresh from off the land driving the ship southwest back across the Bay. The run across the Indian Ocean was one of endless frustration as the trade winds blew light and contrary keeping the ship becalmed for weeks at a time in the southern Indian Ocean. The expected westerly gales off Cape Agulhas were made worse by the shallow and choppy waters encountered as the Star of Greece beat to westward. In desperate need of freshwater and food, Bill Legg decided to put into Ascension Island. The ship lay at anchor in the roads for 48 hours whilst water and supplies were taken aboard.

Departing Ascension for St Helena, the south-easterly trades blowing fresh and strong, those sailors on watch on the night of August 26th were shaken from their duties by a deep and rumbling crack that shook the ship. The very water itself vibrated unnaturally as the evening sky to the east filled with hellish lights. As the sun rose through a sickly orange haze those aboard the Star of Greece could only wonder at which volcano had erupted to the east of Africa. As the days progressed the ship sailed through a sea covered in grey volcanic dust. The fiery orange dawns and sunsets continued for many days as the ship sailed home. Day time temperatures dropped noticeably as the ash from the eruption of Krakatoa filled the sky blocking out the sun. Back in the Dutch East Indies, the eruption and resultant tidal wave removed Anjer and hundreds of other villages and towns from the map. The blast had destroyed most of the island and sent a 130 foot-high tsunami racing across the Indian Ocean, killing more than 36500 people from Indonesia to India.

The Star of Greece passed St Helena soon after reporting 'All well,' and sailed onwards to England. The ship passed the Island of Boa

Star of Greece

Vista on August 6th as she cruised between the Cape Verde Islands and the coast of Africa. From here the trade winds dropped away to naught and having drifted north for several days Captain Legg yet again set course north by west searching for the elusive westerlies. Instead, those aboard the Star of Greece were met by light and unreliable northeast to southeast breezes. The Scillies were passed on September 11th 1883, 142 days from Saugor, with the Downs anchorage being reached three days hence.

Even for the old-time Blackwaller's (frigate built clippers) a trip of 148 days from berth to berth was unusual, much was explained when the Star of Greece was taken into dry dock after discharging her cargo. The weed and barnacles growing below had to be spaded off so thick was the mass. She spent three whole days being cleaned and painted with a fresh coat of patent red lead-based anti-fouling agents. The clipper was shifted back to her East India berth for a general survey and refit, well deserved after 9 months and 24 days at sea.

The Star of Greece was placed on the berth for Calcutta whilst alongside the Star of Russia commanded by John Simpson took on a general cargo for Melbourne. Loading was finalised for the Star of Greece and she was cleared out of Poplar on November 1st 1883. Soon the clipper was towed down to Gravesend to await her time of departure. The ship left the anchorage on the 8th headed down Channel. She passed Dover that afternoon and Beachy Head the following morning and soon after bid farewell to her tug and pilot off of the Isle of Wight. The Star of Greece moved into open waters where she encountered light to moderate southwest breezes until 35° north. Fresh to strong northeast winds emerged in the zone of the trade winds and William Legg made good use of these until reaching 11° north when the winds failed altogether.

The clipper eventually rounded the Cape along the line of 38° to 42° south making her easting run into the Indian Ocean. Captain Legg set a course up through the middle of the Indian Ocean as the ship sailed north by northeast towards Bengal.

Lithograph of the Krakatoa Eruption, circa 1883.
Illustrated London News.

Star of Greece

At 7:00 am on January 21st, at latitude 13.49° south and longitude 85.45° east the clipper passed through several long lines of pumice running from southeast to northwest, a graphic reminder of the Krakatoa eruption several months before. The Star of Greece had entered the ocean at the worst time of the year, at the height of the cyclone season. At least one major storm had already passed east of Mauritius giving more than one vessel a savage dusting. Reunion Island had not been so lucky.

Those aboard the Star of Greece had a rough time of it striking the edge of a cyclone as she crossed the equator for the second time. As the worst of the dirty weather passed the carpenter once again removed one of the hatches to ventilate the hold. Upon lifting one of the boards he was nearly overwhelmed by pungent sulphur fumes, a cloud of steam and smoke billowing up from below. The coal was afire and the man rushed aft to inform Captain Legg, who gave orders for the boards to be replaced and battened down again. The other two holds were checked but no fire was detected. It was here that Harland & Wolffs' ingenious design feature, the three iron bulkheads came into play keeping the fire from spreading to the other two-thirds of the ship's cargo. Then all the ventilators, including those with direct access to the hold were stoppered shut. Wet sails were then lashed down over the hatch covers to further prevent the flow of air into the hold below.

As they sailed north a close watch was kept upon the temperature below via thermometers placed in the 'tween deck spaces. The hull by the seat of the fire began to glow and the paint started to bubble and peel from the heat. Racing north as fast as the wind would carry her the ship was in a desperate situation. The carpenter was forced to drill small holes in the deck above the fire. A fire pump was rigged up, water was poured below via a fire hose. The crew stayed at the bilge pump pushing out seawater as fast as it was pouring in. Bill Legg ordered the boats provisioned and swung out on the davits in preparation for abandoning ship.

The northeast monsoon allowed the ship to cruise comfortably up into the Bay of Bengal away from the worst of the weather. The clipper hove to off the Sandheads pilot station, the fire appearing to have been

Star of Greece

contained to a single hold. The Sand heads pilot brig hove into view on February 12th, 96 days from the Downs. A steam tug was soon hastily contracted to take the clipper up the Hooghly. As the tow line was hooked on William Legg lent over the fo'c'sle head railing and cried out to the tug's skipper, "Go like hell the bloody ship's afire!" The tug put on all possible steam and the race was on. Calcutta was reached on the 14th and the ship was hauled to the river's edge below Garden Reach. With the mooring lines attached a tender came alongside to help put out the fire.

Dense grey clouds of smoke and steam boiled out from below as hundreds of gallons of water were pumped onto the smouldering coal. The Star of Greece settled low in the water as her weight increased by more than 100 tons and in time the fire was extinguished. The clipper was moved up to the coaling wharf to unload her hellish cargo. The Star of Greece was docked and the hold inspected. The shifting boards and sacking used to stow the coal were a mass of charred and soggy cinders whilst the paintwork beside the seat of the fire was a mess. Surprisingly the ship was fundamentally undamaged. She was drydocked for repairs then lay at anchor for more than a month as the crew awaited shipping prices to improve. It was not until the middle of March that Captain Legg was informed that the clipper had been chartered to take on sugar in Mauritius, a destination that the captain had done his level best to avoid at this time of year.

The Star of Greece set sail from Saugor on April 1st with the northeast monsoon blowing fresh across the Bay of Bengal. The clipper ran her course down south by southeast to avoid the African coastal winds. The southeast trades were found to be unreliable making for a slow run to Mauritius. The welcoming beacon of Pointe Aux Canonniers was sighted on the evening of May 11th and the ship hove to in the outer roadstead of Grand River Bay early the next morning, 41 days out from Diamond Harbour. Once within the safe confines of Port Louis, there was a long wait whilst the vessel's assigned load of sugar came in from the mills to the wharf. Over the course of two weeks, the bags of Mauritian sugar were stowed aboard by local lumpers from barges assigned to lighter out the bags which were then hand loaded into the clippers three holds. 1884 was a bumper year for sugar production and it was in high demand right around the world

Star of Greece

with over 125 million kilograms being exported worldwide, 1800 tons of which was to be carried by the Star of Greece.

Loading was finished by June 20th and the vessel was cleared out and moved to the departure buoys. It was another week before the ship was ready to depart as finding enough crew to refill berths vacated by deserting sailors took the local boarding house masters some time on the tiny island. The Star of Greece sailed from the Port Louis roads on June 27th with Captain Legg steering a course northeast, a freshening breeze coming off the starboard beam. Her passage north was made in good time arriving off Saugor Island on July 26th, 29 days from Mauritius.

The Star of Greece was towed up the Hooghly and brought into her moorings to discharge much of her load of sugar. Soon after the ship was moved across to her loading berth to receive a much-reduced cargo of jute. With freight rates at an all-time low, she loaded rice, tea, sacking and linseed as well as her regular cargo of Ralli Brothers jute bales. As the crew were making preparations to get underway that one of them was killed on August 13th 1884. Young Able Seaman, 24-year-old A. Jansen, missed his hold and plummeted into the river whilst setting up the rigging high up on the foremast. Knocked unconscious by the fall the poor unfortunate man was drowned. The body was taken ashore and an inquest held before the seaman Jansen was interred at a local Christian cemetery. With her holds filled with a variety of raw products the Star of Greece departed Calcutta on August 27th and was taken under tow to Diamond Harbour. The pilot came on board and guided the ship out to safety. Farewelling the pilot-brig outside the bar William Legg sailed his ship down into the Bay of Bengal on the 29th bound for London.

Once round the Cape of Good Hope light northwest to southwest winds were the norm, all the way to the trades which soon kicked in strong and steady from the southeast, all the way to just north of the equator. Rolling home across the South Atlantic the Star of Greece signalled 'All well', as she breezed past Ascension Island. Once across the line the ship encountered light airs and calms from the east and northeast in the zone of the northeast trades.

Postcard of Port Louis anchorage with sugar barges in the foreground, circa 1880.

Port Louis, Mauritius, circa 1880s.
Mauritius Library Archives.

Star of Greece

On November 28th at 22°N, 37°W the clipper passed the Swedish ship Edvard, under Captain Akermark, bound for Hull, the two vessels signalling their identities as they passed slowly by. The sluggish winds continued to 27° north, the ship making little headway and becoming increasingly difficult to steer having been so long out of drydock.

Looking out over the hull from the bowsprit, waving green weeds, several feet long could be observed waving lazily in the currents as the clipper slid quietly through the tropical waters west of Cape Verde. Heading north the Star of Greece found herself beating up against freshening north to northwest winds which brought with them bitter arctic rains and rough seas. The further north Captain Legg took his ship the worse the gales became as the winds backed around to the west and then southwest. The Lizard was passed on December 17th 1884, and the Star of Greece hove to under Wolf Rock to take on a pilot and she sailed to Prawle Point that afternoon in search of a tow, 110 days from Sandheads. Three days hence the ship was safely berthed and discharging her cargo at the East India Dock, and Captain William Legg stepped ashore from his beloved ship for the last time.

By 1884 Corry & Co' were moving away from the Indian jute trade and into colonial and global shipping routes. The only clippers within the fleet currently on the berth to Calcutta were the Star of Denmark commanded by John Legg, the Star of Albion under Captain Hughes and the Star of Bengal which came home to London via San Fransisco with a cargo of wheat. The Star of Scotia sailed from Calcutta with jute for New York and was then placed on the berth for Melbourne whilst the Star of Persia set forth from Hull bound for 'Frisco. Corry's Star of Germany sailed out from Gloucester with general cargo and came home via Cape Horn, loaded down to her marks with wheat, the Jane Porter was due home with rice from Burma. The crack clipper Star of Russia also made a run out to Melbourne and came home to Falmouth for orders. She sailed on to Havre to discharge, before returning to London.

1884 was also a year of misfortune for the Star line with a series of mishaps striking many of their vessels both at home and abroad. The Jane Porter was posted missing after 120 days at sea having departed

Star of Greece

Moulmein's Amherst Roads on March 31st 1884. She eventually turned up at Falmouth for orders after 175 days. She had suffered from unusually light winds and contrary gales and was forced to put in to the island of Flores for fresh food her crew suffering severely from the mariner's curse, scurvy. The Star of Germany whilst sailing past the Falkland Islands suffered heavy damage when a series of huge waves swept her decks washing away everything not lashed down, taking out most of the starboard bulwarks. The ill fortune for Corry's clippers continued when the Star of Italy collided with the pilot brig off of Sandheads and even worse was to come when back in London.

The Star of Italy was on the berth for Melbourne taking on general cargo in the West India Docks. A fire was accidentally started by an apprentice in the ships halfdeck house. The deckhouse was gutted, the apprentices losing all of their belongings. The deck was badly charred and much of the cargo ruined by heat and water from fire hoses getting into the 'tween deck space and hold below the fire. A whole suite of sails, as well as glassware, bales of fabric, leather and case goods were lost. The ship's cargo was discharged and the clipper drydocked for some very expensive repairs, of the fate of the hapless apprentice there was no word. That same morning a fire broke out in the basement of Corry's offices in Fenchurch Street, causing extensive damage to the building and its contents. Yet if Corry's thought that their luck could not get any worse they were very wrong. They received word from San Francisco that the Star of Erin had arrived with Captain Coulter dead and most of the crew struck down with gastrointestinal fever. The ship had to be sailed in under the command of the first mate with a skeleton crew as most aboard were too ill to leave their bunks.

Star of Greece

Belaying Pin Soup!

Her cargo discharged, the Star of Greece was laid up for the Christmas and New Year holiday. On January 10th 1885, her new skipper, Captain John Legg, late of the Star of Denmark, had the ship hauled out to the dry dock for a clean and paint. The drydocking was completed by the 13th and the ship was then shifted over to the West India dry dock for a complete refit and survey. With this complete, the ship was hauled back into the South West India Dock and put on the berth for New Zealand, her destination Port Chalmers. The ship's holds were filled with a cargo of iron goods, timber products, leather goods, beer and spirits, drugs and medical equipment, cloth and haberdashery, preserved and tinned foods, cloth and clothing, rugs and carpets, general hardware, floorboards, machinery and farm implements.

The Star of Greece's new master came with a reputation that put off many an experienced shellback from shipping out aboard any vessel in which he was in charge of. John Legg was a 30-year-old Irishman from Carrickfergus. He was a Corry's man from the first time he went to sea having completed his apprenticeship aboard their old wooden clipper Queen of the West, as well as their maiden iron ship the Jane Porter. Finally aboard the crack jute clipper Star of Persia under the guidance of Corry's veteran skipper, Captain John Simpson. He then worked his way up through the company's officer ranks shipping out as a second mate aboard the Star of Denmark. John Legg rose rapidly to become first officer aboard the Star of Greece, under William Shaw, from March 1875 to June 1876. It was under Captain Shaw that he learnt the master's art of driving a ship hard and her crew harder. He did not have the charisma or strength of personality that bred unwavering loyalty in men the way William Shaw did. He was an astute politician and knew exactly how to treat those in higher positions of importance and power than himself. Thus in the eyes of his social betters and more importantly his employers, John Legg could do little wrong.

Star of Greece

Like most of Corry & Co's starting skippers, John Legg's first command was the Jane Porter, which he took charge of in 1877. He quickly established a reputation as a sail carrying passage maker who was not afraid to put a ship and crew through their paces. Each one of his trips for Corry's showed a handsome profit and he was soon given command Star of Denmark. Eventually, he gained command of the Star of Greece. The clipper was well past her glory days by the time John Legg returned to her. She had not made a decent passage since the days of William Shaw and he was determined to turn this record around.

Jack Legg's command style was a far cry from the buccaneering risk-taker that was William J. M. Shaw, who had a well-developed reputation of setting cracking passages no matter which vessel was under his command. In contrast to this William Legg was the complete opposite. He was a man with a generous and jovial nature who loved his creature comforts and had a taste for the finer things in life. He was happiest standing upon the poopdeck of his ship with a glass of decent claret in one hand and fine cigar in the other. A cautious and calculating ship's master William Legg always managed to bring his charges home with little damage and at little extra cost to his employers. He was no great sail carrier or passage maker and the Star of Greece's reputation as a crack clipper had suffered under his care. He was loved and respected by those who served under him and few crew members deserted whilst under his command.

In direct contrast to the famed clipper's previous masters, Captain John Legg came aboard with rather dark and unsavoury reputation amongst British sailors across the globe. Known to all as 'Black Jack' or Bully Legg, Captain Legg was a regular 'Bucko' who drove his crew by the power of his fists, and his general brutality which made life hell on board for the crew. John Legg was a man appointed by Corry's because they could rely on him to drive his crew and ship to the limit, and sometimes beyond, in the quest for a fast passage and quick profit.

In turn, he employed handpicked boatswains and officers who would not spare their voices, fists, or iron belaying pins to keep the crew at work whatever the sea state or weather. Often the usual result was individuals or sometimes entire watches being pushed towards to mutiny because of cruel and unusual punishments, violence against the crew, poor food, horrendous living conditions and exhaustion from being driven too hard. It was a regular occurrence for Black Jack's crews to desert ship even as the anchor was going down thus leaving the officers to auction off the deserting sailor's belongings and keeping their uncollected pays. There were times when those hard-pressed men would fight back and always to their detriment. For even if they managed to get Bully Legg into court there was little chance of conviction for his tyrannous ways or even a fine as an admonishment for his wanton cruelty and violence.

John Legg's reputation for tyranny and for creating discord amongst his crews was well deserved and he would have been at home on an American clipper sailing from London to San Francisco. In 1881 when in command of the Jane Porter on a voyage to Port Adelaide, Captain Legg ended up with much of his crew in prison. Ten members of the crew mutinied and refused to work once the ship had entered port and Black Jack had the whole lot arrested and thrown into irons. In the resultant court case held at the police court in Port Adelaide the ten men; Charles Peterson, Martin Hudson, Benjamin Hadon, Robert Cutting, George Manson, Thomas Hayes, Wilhelm Hanstrom, Albert Lawrence, Peter Nelson and Alfred Olsen all stated that they had been worked too hard in dangerous conditions, the mate and the boatswain had used violence and threats against them (backed up by iron belaying pins, knives, fists, and revolvers), and had treated the men as little more than animals or slaves calling the hapless sailors donkeys, monkeys, and wooden men. Those in court including the magistrate laughed at this. Each man stated that they would prefer to go to prison than again set foot aboard the Jane Porter under Bully Legg's command. The amused police magistrate duly obliged the men, sentencing all ten to four weeks in gaol with hard labour.

This court case was not an isolated incident with stories coming thick and fast of whole crews deserting, of violence and occasionally death as a result of John Legg's viciousness. According to some sailors, hardly a day passed but one of them was beaten and ill-used by Black Jack's officers or by the Captain himself. In his later career there emerged even more stories where John Legg ended up before a magistrate being accused of cruelty and violence against various crew members who had finally refused to take any more and had rebelled against his tyrannical ways. In 1892 as skipper of the Star of Russia on a voyage from Newcastle to 'Frisco 'Black Jack' Legg had to deal with a mutiny 14 days from port when most of the crew refused to come out from the fo'c'sle to work. The officers set to with a will, beating, and threatening the crew with seven forms of hell if they did not get back to work. To emphasise the point John Legg wielding a large calibre pistol chased the ring leader, the boatswain, a man named Frank Butler Harwood, all the way to the fo'c'sle threatening to shoot him if he did not come out and go aft to the cabin. To preserve his own life, Frank Butler, as he liked to be called, allowed himself to be placed in irons and led into captivity.

On the return trip from San Francisco to Liverpool members of the crew again mutinied and ended up before a Queenstown magistrate for their troubles. Despite overwhelming evidence attesting to the sociopathic treatment of the men by Captain Legg, the offending sailors; Thomas Bell, B. Wilson, Charles Long, Jeremiah Crowley and C. Collins were sentenced to one month in prison with hard labour and were also fine two days wages to cover costs. Another seaman, Julius Grower accused Captain Legg of striking him in the head with a leather-bound iron shot. Along with the oversized revolver, this handy little tool was Jack Legg's signature weapon, after all, he was not called 'Black Jack' Legg for nothing. Needless to say, the charge was thrown out and the sailor laughed at by the magistrate who complimented John Legg for his conduct and bearing as a model ship's master.

Having finished loading the Star of Greece was hauled out of the South West India Dock on February 9th and moved down to Gravesend. A tow was arranged early the next morning and she took her departure from Beachy Head. A news article published in the Otago Daily Times dated May 8th described the clippers' journey in great detail;

"Captain Legg reports leaving London on February 9th by tug down the Channel until she reached Beachy Head light at 4:00 pm of the 10th she made sail and cast off the pilot boat. She had westerly winds

Star of Greece

down channel and took her departure from Start Point on February 15th; met moderately fine weather across the Bay of Biscay (where she was spoken on February 18th), and picked up the N.E. trades on March 1st in latitude 27N, longitude 20W; experienced very light trades, which carried down to latitude 2N, longitude 26W, on March 10, and were followed by light variable winds and calms; crossed the equator on March 12, in longitude 27W, and took up the S.E. trades in latitude 1.3°S on the same day, they proved very light and carried her down to 16°S and longitude 33.3°W on March 19 and thence she had light variable winds until the 22nd, followed by E to E.N.E winds until march 26, then in latitude 30°S, longitude 25°W, when it veered to N and N.E., backing to NW, with light weather until April 3 in latitude 43°S, longitude 5.3°E, the meridian of Greenwich having been crossed the previous day in latitude 43°S, still keeping variable winds from N.N.E. to N.W., she rounded Cape of Good Hope on April 9; thence she had S.E. to E.S.E and southerly winds for two days, and afterwards took the passage winds. They were moderate and very unsteady, veering west to S.W., all across the Southern Ocean taking her 35 miles north of the Crozet Group. The meridian of Cape Leeuwin was crossed on April 23 in latitude 48.3°S; and still keeping westerly winds she passed the island of Tasmania on May 2, in latitude 48°S, thence she had strong winds from N to W and S.W., and made the Snares at 4 pm of the 6th inst., had fresh S.W. winds up to the morning of yesterday, and afterwards carried moderate westerly winds along the coast. Her easting was made in the mean parallel of 48°S, and neither ice nor wreckage was seen in the Southern Ocean."

Otago Daily Times, Issue 7246, 8 May 1885, Page 2

The Star of Greece hove-to off Cape Sanders at 4:15 pm on May 7th, 81 days from Start Point. She anchored for the night outside the Heads to await a tug that would tow her into Port Chalmers the next morning. Contracted to the New Zealand Shipping Company, the Star of Greece was towed through the Heads and down to the inner harbour by the tug S.S. Plucky. The customs tug was met at the entrance to the inner port anchorage and after receiving customs clearance and pratique the ship was hauled into Bowen Pier to discharge her 2600 tons of cargo.

Star of Greece

Port Chalmers circa 1870's. Photographer: David De Maus.
National Library of New Zealand.

Star of Greece

The Star of Greece had her cargo discharged by the end of May and began taking on several hundred tons of rock ballast in preparation for a trip across the Tasman Sea to Newcastle N.S.W., where Captain Legg hoped to take on a load of coal for Calcutta. The ship was cleared out on June 16th in preparation for her departure from Dunedin's sheltered waters. Captain Legg had his ship towed back out through the Heads and after farewelling the pilot boat set all plain sail on June 19th bound for Cooks Strait. She reached this treacherous stretch of water on June 22nd pushed along by freshening southerly winds.

It was as she approached the Strait that Black Jack's antics came back to bite him. He had been riding the crew hard since before the Star of Greece had left Port Chalmers and they were in a rebellious mood. A particularly angry sailor named Read had refused an order from the first mate, John Stevenson. This act of open defiance was reported to Captain Legg who in turn summoned Able Seaman Read to the poopdeck to explain himself. Black Jack had perhaps been focussing too much meanness upon the young man for two of his friends, a couple of sea lawyers named Edward Hayward and Edward Paull, followed him up to the quarter-deck, both watches following close behind.

Upon their arrival, one of the belligerent sailors asked the Captain what he wanted from Read. Seething with anger Bully Legg rushed down from the poop grabbing Hayward by the throat attempting to choke him. The two men struggled and in the ensuing affray, John Legg became entangled in some ropes scattered about the deck and tripped, Hayward falling on top of him. Lifted off the Captain by some of the crew Hayward was escorted back to the fo'c'sle as the first officer attempted to bring calm to the situation. The rest of the crew including seamen Read and Paull were looking on with great interest, the fate of Black Jack and his ship hanging in the balance. Captain Legg took himself off to his cabin and returned a few minutes later armed with a large revolver. Storming off down the deck he came to the entrance to the fo'c'sle and waving the pistol around, threatened to shoot Edward Hayward if he did not come out and take himself down to the cabin to be placed under arrest.

Loading coal at Newcastle, N.S.W. Circa 1880's.
Photograph by George Freeman.

Star of Greece

Newcastle Co-operative Steam Tug Company's Steam Tug 'Goolwa', circa 1880s.
www.ferriesofsydney.com

Star of Greece

Hayward emerged in a sudden rush trying to disarm Bully Legg as he and the ship's officers attempted to place iron manacles upon Hayward's arms and legs. Black Jack grabbed the unfortunate sailor's fingers, bending them back to the point of almost breaking. Finally contained, the outraged sailor was bundled aft and placed into a storeroom to await his fate. He was later joined by his friend Edward Paull who had attacked the mate John Stevenson during the fracas.

With a mutinous crew successfully cowed and the ring leaders in chains, there was little trouble for the remainder of the voyage. The Star of Greece reached the roadstead off Newcastle on July 6th, 17 days out, pilot to pilot. The clipper was towed into Newcastle harbour later that afternoon by the steam tug Goolwa and taken to the arrival buoys to await her load of coal. Upon arrival in port Captain Legg had the police launch summoned and the two recalcitrant sailors arrested and taken ashore to be charged with failure to follow lawful orders and assault.

At the subsequent trial in the Newcastle Magistrates Court, the two sailors Hayward and Paull were tried for their apparent crimes. Ten witnesses were called, the first mate, the steward and the two apprentices stood by Captain Legg whilst the rest of the crew, of all ranks from the second mate down, testified in favour of the defendants. Despite the abundance of witnesses speaking in defence of the two seamen, the local magistrate preferred John Legg's version of events. Hayward was found guilty of assaulting his captain and was sentenced to ten weeks gaol with hard labour. Edward Paull was found guilty of disobeying a direct order from his captain and was sentenced to five weeks in the local clink with hard labour. An incredulous crew stalked out of the courtroom refusing to sign on again as long as Black Jack was in command.

The Star of Greece was scheduled to load coal and was anchored in the river awaiting her turn at the coal loading wharf. It soon became apparent to Captain Legg and the ship's agents that the cost of the coal and shipping it would mean that the trip to Calcutta would force the ship to run at a huge loss.

Star of Greece

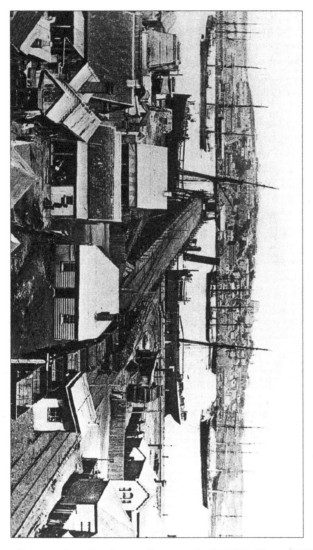

Loading coal at Stockton, Newcastle N.S.W. Circa 1880s.
Newcastle Herald Archives.

Star of Greece

So with freight rates being so bad this year it was decided in London to send the Star of Greece to Bengal in ballast instead. The clipper was moved out to the north harbour buoy on July 13th. The taking on of ballast was completed on the 16th as the customs cutter came alongside. After gaining his clearances Captain Legg had his ship cleared out and made ready to put to sea.

The boarding house runners of Stockton were still rounding up a few sailors for the ship's crew. It was not too long before the fo'c'sle was again filled with drunk, desperate, unwary or case-hardened sailors willing to sail under Bully Legg's command. July 18th 1885 dawned overcast with light rain, the last sailors taken on were awoken with sore heads by the sharp commands of the boatswain. Turning to they stumbled out of the fo'c'sle and set to work making ready to get underway.

The ship managed to avoid the worst of the tropical storms and no cyclones troubled her passage. The Star of Greece hove-to on August 29th 1885, 42 days out. With the pilot aboard Jack Legg allowed him to guide his vessel through the ever-shifting channels of the Long Sands and into the Saugor roadstead where the ship was to anchor for the night. The tug was met early the next morning and the two craft proceeded upriver on the rising tide. The anchorage of Garden Reach was finally obtained on August 31st, the banks of the Hooghly crowded with vessels awaiting the chance to load.

Captain Legg's angry heart sank when he realised just what this crowded anchorage indicated. Shipping rates were low. Most vessels present had the look of having been laid up for many weeks or months their yards down and sails off. All had awnings spread over poop and quarter decks to shelter crews from the sun and rain of the summer monsoon. As per the usual routine, an apprentice was tasked with scrambling aloft to the mainmast truck carrying the ship's famed 'Golden Rooster'. It was the young lads' task to fasten the gamebird to the weather pennant. However, on this occasion, the brass ornament slipped from his grasp and plummeted to the deck. Crying 'Watch out below!', the luckless boy watched in horror as the lump of metal slammed into the boards at Captain Legg's feet. A startled and then outraged Black Jack looked upwards, murder in his eyes.

Star of Greece

After roaring his displeasure at the quailing lad he picked up the mangled rooster and hurled it into the Hooghly. The rest of the crew set about their duties careful to avoid their captain's wrathful stare. As for the apprentice, he spent his time ashore in the military infirmary, sporting black and purple bruises from head to toe. Upon reaching shore the rest of crew of the Star of Greece were shocked to see the state of many of the buildings along Calcutta's substantial waterfront.

On July 14th at 6:20 am, the city had been struck by a violent earthquake. Undulating waves of movement rolled across the city threatening to collapse many a structure as a series of high crested swells pushed upriver sending those aboard moored vessels into a blind panic. The city escaped major damage as the six-minute-long quake did little more than crack many of the older homes and public buildings. Yet despite the scare, life in Calcutta continued apace as more and more vessels from across the globe lined up to take advantage of Bengals booming markets.

Economic conditions in Calcutta had changed in the last five years as jute exporters like Ralli Brothers shipped their raw products to increasingly diverse markets all over the world. Indian jute was being exported in great volumes to mills in Germany, France, Burma, Singapore and Malaya, Australia, New Zealand, South Africa, Egypt, Turkey, Greece, Spain and ports on both sides of the United States of America. With places like Dundee, London and Liverpool no longer having exclusive access to Bengal's jute exports competition and the increasing use of large steamships to tranship cargos combined to push freights to an all-time low. For the first time, rough jute goods were put onto the market in direct competition with products from Dundee. The mill operators there were forced to specialise in fine wares or go under as the market was flooded with cheap jute-based products from across the world. Those aboard the Star of Greece were forced to wait for shipping rates to increase, or for jute prices to become competitive enough to make loading up and shipping her usual 8500 jute bales worthwhile.

As they were laid up in the Hooghly another Corry clipper, the Star of Albion was inward bound with a load of Cardiff coal. She entered

119

Star of Greece

the Bay of Bengal just behind a building tropical storm that was pushing north towards Calcutta. The monster cyclone swept into the Ganges Delta on the evening of September 21st 1885 bringing with it ferocious winds, torrential waves and an enormous storm surge that pushed well up the Hooghly delta and surrounds. Giant waves washed over many of the islands and lowlands flooding more than 3500 square miles of countryside to the south of Calcutta. The cyclone remained stationary for several days bringing catastrophic flooding further inland and to the city itself. Vessels large and small were smashed to flinders up and down the Hooghly and its tributaries with thousands of people drowned. One of the hardest-hit places was False Point a shoreside settlement near Kendrapada. The gigantic storm waves made a clean sweep of the sandy point carrying away every building and more than 300 souls into the boiling waters beyond. Out in the approaches to the Hooghly, several deep-water vessels came to grief as they attempted to seek shelter.

A New Zealand based paper, The Colonist, published a graphic description of the storm and its effects.

"On Sept. 22 a terrible cyclone occurred at Bengal. One telegram with regard to it states:-"A serious loss of life is reported from False Point. The storm burst with unprecedented fury at Hookeytollah. The port officer, Captain Douglas, his wife and three children, his assistant, Mr Minos, Mr Walker; the agent to Messrs Bullock Brothers, and about a dozen Customs and preventive officers were swept away. The captain of the Tewkesbury, who had gone ashore on business with a boat's crew of six men shared the same fate. At Pattimimdi, the Government offices were levelled to the ground. Not a building, hut or tree escaped. There has been a severe loss of cattle, but few men have been drowned. At False Point several vessels were completely wrecked and others stranded. The barque General Fytche, of Moulmein, was blown five miles inland. The barque Nucotra, of Ava, is a complete wreck and is breaking up fast. The captain and crew are ashore. In addition to the loss of the captain and Lascar crew of the Tewkesbury, the chief and second officers are missing. The European portion of the crew are cruising about the Bay of Bengal in a patched-up boat. Ashore the results are far more serious - 300 Natives were killed. With the exception of a workman and

Star of Greece

lighthouse keeper, there is not a single European left to tell the tale of disaster.

The carcases of thousands of cows, deers, buffaloes, and tigers line the banks, and the atmosphere is thoroughly noxious. The ship Merchantman, bound for Mauritius with a cargo of rice and carrying native passengers, foundered in the Hooghly on Tuesday, September 22. During a hurricane, her sails were torn to ribbons. Soon after she shipped a heavy sea, which broke the forepart of the poop, and carried away her steering gear, and then a succession of heavy swells washed her fore and aft till she went down. The chief mate, D. Henderson and a Lascar, after being ten hours in the water clinging to a lifebuoy; were rescued by the steamer Britannia. The remainder of the 69 men all went down...

A large proportion of the population in the flooded districts has been brought to the verge of ruin by disaster which they could neither foresee nor avoid. The most harrowing reports of distress and misery are received from the afflicted districts, and it is evident that nothing but a vigorous effort on the part of the public can prevent a terrible loss of human life. It is estimated that 3000 square miles of country, with: a population of close upon 2,000,000, are affected by the floods. , The average depth of water is said to be from 6ft to 8ft."
THE CYCLONE IN THE BAY OF BENGAL. Colonist, Volume XXVIII, Issue 4308, 25 November 1885, Page 3

Approaching the mouth of the Hooghly in the wake of such a catastrophe was the 1000 ton jute clipper Star of Albion out of Cardiff, commanded by Captain John McIlroy. Having encountered storm clouds for several days it had not been possible for Captain McIlroy to obtain a position fix and thus without having taken regular soundings nor checked his charts, he inadvertently sailed his ship to the west of Long Sands. Realising his mistake McIlroy gave the order to put about in hopes of sighting the pilot brig. In high winds and a confused cross-sea, the Star of Albion grounded heavily in the murky silt-filled floodwaters as she crossed the shallows east of Long Sands shoal. Such was the force of the impact several bow plates below the waterline were bent and seams sprung.

Star of Greece

It was not long before the forward hold completely filled with water as waves continually swept the ship. The crew realised that their vessel had been bilged and being unable to get her off they abandoned ship. The captain and crew knew beyond doubt that the Star of Albion would soon go to pieces such was the pounding she was taking. All onboard safely reached shore in the ship's boats and soon after were taken upriver by steamer to Calcutta where they were sent to the overcrowded Seaman's Mission and housed as distressed sailors.

Carefully moored the Star of Greece along with dozens of other vessels rode out the tempest in relative safety despite local flooding and storm damage. The cyclone wiped out thousands of acres of jute and other commercial crops. This greatly delayed the clippers' departure as damage to inland river wharves, bridges, roads and railway lines prevented the transport of jute to Calcutta for processing and shipment. The hold was not filled until November 14th and Captain Legg obtained the ships custom clearances the next day. After a careful tow back down the Hooghly, the Star of Greece set sail from Sandheads pilot station on November 20th 1885 homeward bound for London. Jack Legg was not hopeful of a quick run home so fouled was the clipper's bottom with barnacles and weed. With the cooling monsoon winds coming weakly from the north Captain Legg took the ship south-west cruising along the Coromandel Coast in search of the available offshore breezes.

Snow and hail showers battered those aloft as the clipper sailed past the Lizard in the thickening fog. This was followed by heavy rains that swept in with gales that cleared away the worst of the freezing weather as the Star of Greece was heaving to within sight of St Catherine's Lighthouse, Isle of Wight on March 19th 1886, 119 days from Saugor. The pilot cutter put out from Ventnor and was soon followed by a steam tug anxious to take the clipper in tow. Anchoring within the sheltered waters off Shanklin Chine the Star of Greece waited for the storm to pass before setting out behind the tug for London.

Skies quickly cleared as the ship and her tug, the Victor, set sail from Shanklin the next morning making Deal late in the afternoon. Millwall was reached mid-morning on March 21st and the Star of

Star of Greece

Greece was shifted into her East India Dock berth to unload. The unfortunate crew members who had incurred Bully Legg's wrath jumped ship at Gravesend leaving a scratch crew of old men and boys to bring the ship in through the lock. Upon inspection, everyone ashore was shocked and dismayed at the state of the clipper. Her rigging was strained, paint chipped, her hull rusty and bright-work unpolished, but most telling of all was the appalling state of her bottom, covered in weeds more than a meter long in places.

After the jute was discharged Corry's agent had the Star of Greece taken over to the East India graving dock for a complete overhaul. At the same time, Captain John Legg received a dressing down and warning over the state of his vessel. The ship was laid up on the stocks for more than a fortnight as her hull was given a full clean, inspection and paint. Her rigging and sails were overhauled and then the ship underwent a full refit. She left the docks on April 13th and once back in the East India Dock was put on the berth for Calcutta.

The Star of Greece passed Deal light early on May 2nd 1886, in tow and with light south-easterly winds, clear frosty skies and smooth seas she left her tow off the Isle of Wight and sailed serenely down the Channel. The south-easterlies continued increasing in force becoming squally as the clipper rounded Ushant to cross the Bay of Biscay. Jack Legg hung out all his heavy weather canvas to clear the treacherous waters as quickly as possible. The winds hauled around to the northwest holding light to moderate until the northeast trade winds were fallen in with during the last week of May. Light and warm breezes blew off the land until they disappeared altogether just north of the equator.

Entering the higher latitudes, the Star of Greece continued her southing powered along by gale-force southwest winds, passing below the Cape of Good Hope in the first week of July below 44° south. Captain Legg began his easting run across the southern Indian Ocean and up into the warmer climes of the tropics in search of the southeast trades. Light southeast to northeast winds greeted the clipper as she passed the line of 25° south. The winds held light and variable until the ship entered the doldrums southeast of Ceylon. The southwest monsoon with its heavy squalls and constant rains allowed

123

Star of Greece

John Legg to make up time and the pilot brig was signalled on July 29th 1886, 88 days pilot to pilot.

"Pilot brig at the Sand Heads" Bay of Bengal, tinted stone lithograph, circa 1850.
National Maritime Museum.

Upon discharge of her inward coal cargo, the Star of Greece was hauled back to her riverside berth to await a charter. The southwest monsoon continued to bring torrential rains and the ever-present heat and humidity that made life rather uncomfortable for those on ship keeping watch. Eventually, Corry's Calcutta agents managed to secure a sugar charter and the Star of Greece left Garden Reach, behind a side-wheeler on September 21st 1886, amidst violent, swirling winds and rains, the tail end of yet another cyclone. This particularly nasty storm had passed south of Saugor the previous day before heading west. Her passengers, several hundred indentured Bengali labourers destined for the sugar-cane fields of Mauritius, were crammed in the 'tween deck spaces during all of this time. Captain Legg ignored their cries and pleas to be let up on deck despite the stiflingly hot and squalid conditions below.

Star of Greece

Port Louis roadstead was reached on October 26th, 33 days out and after gaining her customs and pratique clearances the clipper was towed between Forts William and George, to the inner anchorage there to await a berth at the loading wharf. Very soon several lighters were rowed out to collect the labourers who were then transported to an immigration depot for further health checks. After shifting enough ballast the ship was warped into the pier beside the loading derricks to await the arrival of the sugar from the Port Louis sugar mills. The Star of Greece was four weeks in port before the last bag of sugar was stowed. Departure was taken from Port Louis on November 23rd soon after the ship was making good time through the storms and calms of the Indian Ocean with the wind constantly on the starboard beam. Corry's latest vessel the Star of Austria followed the Star of Greece a week later headed for Calcutta. Upon entering the Bay of Bengal Bully Legg's ship encountered freshening northerlies that brought with them the first hints of cooling airs from the far-off Himalayas. Christmas Day was spent beating up the Coromandel Coast and the Long Sands were not sighted until December 29th 1886. New Year's Eve was spent in the Garden Reach when the clipper was warped into discharge her 1800 tons of sugar. The Star of Austria did not arrive back in Calcutta until January 23rd, having been caught by dirty weather and then frustratingly contrary winds, 54 days from Port Louis.

Four days after the Star of Austria arrived, the Star of Greece was entering out, her hold filled with a mixture of goods; rice, tea, hessian sacking, cotton, exotic Indian wares and several hundred tons of jute both milled and raw. The Star of Greece charged on down the Coromandel Coast and onwards to Ceylon ahead of a freshening monsoon breeze. The fine weather continued to the equator before the northeast trades were fallen in with just past Madeira. Sailing into the North Atlantic the Star of Greece met with the usual dirty weather and boisterous seas as she approached the English Channel. Moving up from the southwest the clipper was met by a series of northeast to northwest gales that frustrated Bully Legg's chances of making quick time into the Thames.

Star of Greece

The Lizard was reached on May 13th, as the ship hove to awaiting the appearance of the pilot schooner, 104 days pilot to pilot. The tug was picked up off the Isle of Wight on the 15th as the ship thrashed her way up the channel into blustery headwinds and high seas. Deal was made the next day and the Star of Greece was taken up to her Gravesend mooring to wait out the adverse weather. There Captain John Legg left the ship to take up command of another of Corry's clippers. Taking his place was Captain Henry Russell Harrower, unfortunately for Corry's the intrepid former master of the Star of Persia and Jane Porter was still at sea, hurrying back from India aboard a P&O steamer bound for Brindisi, Italy.

Captain John Legg in retirement.
Legg family Archives.

Star of Greece

Dreams of Adventure

Captain Henry Russell Harrower was born on February 6th 1859 at Broughty Ferry near the city of Dundee. He was the 5th son of George Kerr Harrower, church deacon, grain and hay merchant, and insurance broker, and Jemima Margaret Wright. The family lived in a large house at Yew Bank, Monifieth Road, Broughty Ferry. Henry's siblings grew up listening to the tales of their grandfather's overseas adventures, an aspiration which their father actively discouraged. Despite George Harrower's best efforts to steer his children, particularly his sons into money and business, they were afflicted with wanderlust and thirst for travel that would never be quenched or satisfied by the stifling social conventions of Calvinist Dundee society. The children; Alice (b.1852), George (b.1853), Frank (b.1856), Robert (Rob or Bobby) (b.1858), Henry Russell (or Russell to his friends) (b.1859), William (Bill) (b.1861), Patrick (Pat) (b.1863), Mary (b. 1866), and Tom (b. 1868), James (b.1867 - d.1870), were almost all affected in some way by the influence of their grandfather's example. Captain George K Harrower[snr] had been a swashbuckling, womaniser who was profligate with his and other people's money. Sailing the seas as a master aboard various East India Company clippers and frigates Captain Harrower developed a rather notorious reputation being convicted of bigamy and eventually declared bankrupt. His son George Junior was determined that none of his children would follow in his wastrel of a fathers' footsteps.

He encouraged his sons to move into the finance and banking sector, helping his offspring to gain employment as bank clerks for George, Frank, and Robert. Alice, the eldest was afflicted with wanderlust first. She married early and migrated to Adelaide, South Australia with her husband James Bishop, despite her father's wishes that she should stay to care for her parents in their old age. Robert became a successful banker in London and died in 1907. Frank, despite being safely ensconced in a well-paying bank job, was soon enough getting itchy feet. He became a merchant banker travelling to Europe, the Middle East and Africa as part of his job. He died of fever in Zanzibar in 1891, aged 35, having worked there as a bank manager. Robert was travelling from an early age, his work as a merchant banker frequently took him to Calcutta and Bombay. It was whilst visiting his wife's family in West Bengal that Robert developed a severe fever. The family decided to take a steamer back to London. The stricken Robert never made it home, dying of Tuberculosis on April 29th 1891. He was buried at sea aged just 33. Patrick Harrower followed his brothers starting life as a clerk and eventually moved to London becoming a highly successful shipping and insurance agent. He died in 1936, aged 73. Mary married James Marr a mercantile clerk, moved to Middlesex and raised two children. Thomas and James never made it to maturity both passing away before the age of 5.

Three people had the greatest influence on the life of Russell Harrower. His grandfather, whose many adventures been retold countless time, filled Russell's head with dreams of adventure in faraway lands. Then there was Alice, his eldest sister who looked out for the boy when his elder brothers picked on him. She was very much his daily carer when their mother was busy with the running of a large and active household. Their parting when she married was sad enough, but the separation was made worse when Alice migrated to Australia. Despite the distance, the two remained incredibly close, an attachment that was to have terrible consequences a few years after Russell's death. The last and perhaps most important person in Russell's life was his younger brother William. Less than two years separated the brothers and they were inseparable, each sharing the others love for adventure, stories of their Grandfather's escapades, and the sea.

As a young lad, Henry Russell Harrower was determined to follow in his grandfather's footsteps. His father George was determined that if this was to be the case then it would be in the employ of a company with sound prospects. Thus, he managed through his contacts to gain Henry a position as an apprentice aboard a clipper barque owned by Imrie, Tomlinson & Co. His maiden trip abroad would be upon the 550 ton Tamaya, under Captain Halliday. The vessel was put onto the berth for the West Coast of South America in the nitrate trade. Henry boarded the Tamaya in October of 1873 and spent the next nine months learning the ways of life at sea aboard a deep-water clipper.

Photograph of a painting of 570 the barque Tamaya.
National Maritime Museum.

Star of Greece

This voyage was not to his liking and he returned home disgruntled and scarred. He was not to be deterred and had his indenture papers transferred to a brand new jute clipper, the 1900 ton St Enoch. This made to order vessel was built by Dobie & Co, for the Forfarshire Clipper line, owned by W. S. Croudace of Dundee and put on the berth to Calcutta. Henry sailed as an apprentice in October of 1874 under Captain Ovenstone. The ship made two runs out to India and back with Henry impressing his skipper with his abilities. Upon the St Enoch's arrival back in Dundee in April of 1876 Henry left Captain Ovenstone's service only to be replaced by his younger brother William. The 15-year-old was excited to follow Henry's lead and the experienced Ovenstone was happy to have him aboard.

After two months ashore Henry Harrower took to sea again, this time signing on aboard the Shaw, Savill Co' owned ship, St Leonards bound for New Zealand with migrants and general cargo. The 1054 ton clipper commanded by Captain Richard Todd, sailed out from England with 51 migrants, bound from London to Wellington, making the trip in 106 days. Henry spent Christmas and the New Year in New Zealand before the ship sailed for home in early January, arriving back in London on April 23rd 1877. He left the plucky little main skysail yarder for a run ashore which turned into an extended stay of three and a half months. Henry's father must have wondered if his son would ever stay with anything so great was his wanderlust. He had already squandered two chances at an apprenticeship and had instead signed on as an Ordinary Seaman on a voyage to the colonies.

Harrower's next run involved a certain amount of luck as he was signed on as a watchkeeping 3rd mate aboard the 1200 ton Malabar, a wooden, frigate built vessel belonging to Green's Blackwall Line. The smart little Blackwaller set sail from Gravesend on August 11th 1877 bound for Madras. The barque had a rather light air passage and did not arrive at Madras until December 8th, after 119 days at sea. From Madras, the Malabar sailed back to Natal before returning to Madras to load for Colombo and thence back to Calcutta. Once in Calcutta Henry gained his Coastal Second Mates ticket on October 19th 1878. Henry Harrower, or Russell as he preferred, decided to stay with the ship and was immediately promoted officially to the position of second mate.

Star of Greece

**Jute Clipper, St Enoch of Dundee, 1941 tons, owned by the
Forfarshire Line, circa 1875.**
Dundee Archives, Dundee Public Library.

Star of Greece

Blackwall Frigate MALABAR, 1219 tons, built 1860 at Sunderland.
Lithograph by T.G. Dutton, National Maritime Museum, Greenwich.

Star of Greece

After running back and forth around the Indian Ocean the Malabar, at last, arrived off Deal from Barbados on June 9th bound for London. She reached Blackwall the next afternoon and Russel Harrower took his leave from the barque after 22 months at sea. With a good conduct discharge from his captain and a letter testifying to his seamanship skills, Russell headed to the home of his brother at Picton Villas in Highgate London for a much-needed rest. He had plans to finally gain his deep water Second Mate ticket and needed time to study and put in his application at the St Katherine's offices of the Board of Trade. There was tragic news waiting for him upon his arrival, news which was to shape his life and thinking for some time to come.

William Harrower's ship, the St Enoch set sail on March 22nd 1878, from Dundee outbound to Bombay with 2500 tons of coal aboard. The clipper departed in company with the ships Jason and Culzean both similarly laden and headed for the same destination. Of the three vessels, only the St Enoch failed to arrive. It was thought that the ship was delayed, her bottom fouled with weed. Initially insured for £28000 her premiums went up by several guineas per ton the longer she stayed out. Eventually, the clipper was listed as missing by Lloyds as relatives and friends of those aboard the St Enoch waited anxiously for news.

It was months before anything at all was heard and it was scanty news at best. A chest of letters from a seaman aboard the ship washed up in the middle of Quiberon Bay, France, but there was little to indicate how it arrived there. Amongst the 33 missing sailors aboard were three young friends, William Burness 19, William Harrower 18, and James McMaster 14, all from Broughty Ferry. These young men had all joined the St Enoch as apprentices under Captain William Browse. It was not until June 1878 that news finally arrived, hinting at the likely fate of the missing ship. Aside from the box of letters being found near Auray, the wreckage of a ship's boat with St Enoch inscribed in blacks letters on one of the larger pieces was washed up near St Gildas, along with seven boxes of preserved meats and other wreckage. At Belle Isle another ship's boat came ashore, this time intact, also with the name of the St Enoch painted upon her transom. There could be little doubt now that the missing clipper had gone

Star of Greece

down with all hands somewhere off the coast of France, within sight of Quiberon Bay.

What was known was that the St Enoch had sailed from Dundee with the wind freshening from the northwest and the glass dropping steadily from 30.19 to 29.18. Passing Deal on the 24th of March, the cool dry conditions continued along with the smooth seas, with the wind veering to the north. She was then seen to sail southwest around Ushant and enter the Bay of Biscay, from there the clipper was never sighted again. The breeze at the time the St Enoch passed Ushant was cold and dry. Winds were fresh to strong from the west-northwest with occasional gusts and squalls. Seas were slight and visibility fair in the hazy and overcast conditions. The ship and her crew were given up for dead and insurances paid out. The speculation was that the ship had been rammed in the night as she passed west of Belle Isle and the accident had not been reported or both vessels had foundered and sunk.

Russell sat his Seamanship and Navigation exams on July 14th and passed easily. Approval for his deep water Second Officers Certificate was given the next day and he stopped by St Katherine's dock to pick up his papers on his way back to the Malabar where he would continue as an officer aboard the well-travelled Blackwaller. The barque set sail on July 18th with coal, bound for Demerara to take up a load of sugar. From South America, she sailed for New Brunswick to discharge before taking on a cargo bound for Marseilles. The Malabar was kept in the sugar and timber trades sailing between Central America, the Caribbean, French Canada and France for the rest of Russell Harrower's time aboard. He next set foot on English soil on September 7th 1880 having just put into Deal from Barbados. Russell immediately set off for London and his brother's home in Highgate where he set about catching up on his studies for the next round of Board of Trade exams. Russell again passed with flying colours earning his First Officers Certificate on September 15th almost immediately re-joining his ship as her first mate. The barques' next run out was to Calcutta where she arrived in early January to pick up a load of jute for Dundee.

Star of Greece

Once in India Russell learned that the Malabar was being taken out of service, her hull waterlogged, timbers rotten and leaking. He was lured away from life under sail by agents for the British India Steam Navigation Company to take up a position as Fourth Mate aboard the ss Ellora, (Captain J W Seaward), on March 9th 1881. He bounced around various British India steamers, his next appointment aboard the ss Oriental (Captain E C Russell), being cut short after just 15 days when he was dismissed for not having the correct paperwork. This was soon proven to be an error on the company's part and Harrower was given a berth as third officer aboard the ss Busheer (Captain J James), a position he held for 12 months. His last stint with B.I.S.N was as second officer aboard the ss Madras (Captain H Harris). Soon the life as a steamship officer sailing around the coast of India began to pall. He applied for and was given leave to undertake studies for his Ships Masters certificate yet without enough sea time as a first mate the Mercantile Board in Calcutta refused his application. Determined to gain his ticket Russell handed in his notice to the British India Steam Navigation Company office in Calcutta on November 10th 1882 and signed on aboard a homeward bound jute clipper the very next day. The Malabar after a survey by Lloyd's assessor was soon condemned and her hull was broken up for her metal fittings, spars, anchors, cables and deck gear.

Russell Harrower reported aboard the 2000 ton steel clipper Dunstaffnage, he was referred by first officer James Reid to the captain James Welburn. Captain Welburn after inspecting Russell's paperwork which included testimonials attesting to his skills and good conduct was more than happy to welcome a fellow Scotsman aboard his ship for the trip home. The vessel was on just her second voyage when she took up her tow down the Hooghly on November 11th, taking her departure from Sandheads on the 13th. The Dunstaffnage had a very good run down towards Mauritius with the southerly monsoon giving way to the cool winds from the north. The trades proved light and fickle though Captain Welburn still managed to make good time. On approaching Cape Agulhas, the weather turned thick and nasty with gales and high cross seas being encountered 44 days out, between Agulhas Banks and the Cape of Good Hope. Conditions improved dramatically as the ship rounded the Cape of Good Hope. From there until north of the line those aboard the Dunstaffnage enjoyed a fair-weather passage.

The clipper took a further dusting as she passed the Azores, being repeatedly pushed over onto her beam ends and losing a foresail as she battled against an easterly squall. Despite the poor weather, the Dunstaffnage reached the Lizard on February 16th 1883, 95 days from Saugor. On her way up the English Channel the ship came into collision with a fishing smack which foundered, the ship was hove to and the fishermen rescued. Captain Welburn picked up a tow off Dover which then hauled the clipper north to Dundee, where she dropped anchor on the 21st. The jute trade to Dundee was experiencing a boom with twenty East Indiamen moored in Dundee harbour. Harrower stayed aboard just long enough to attend the wedding of his new friend, first officer James Reid to Helen Davie, which took place at Watson Terrace, Dundee. With the celebrations over, Russell failed to farewell Captain Welburn as the skipper had declined to comment on Russell's abilities at sea on his discharge ticket indicating some doubt on James Welburn's part as to the abilities of his second mate. Russell did say goodbye to the remaining officers and crew of the Dunstaffnage before heading home to London.

From Dundee, Russell caught the train back to Highgate. He spent the week before taking his exams in hard study before fronting the Mercantile Offices. There he arrived to lodge his application to sit the triple exams required for the Ordinary Master's Certificate. Fronting up on the morning of Sunday, March 4th 1883 he entered the side entrance of the Board's Examination Office and presented himself to the clerk who directed him to the foyer outside the medical officer's rooms. Soon after the Chief Surgeon called Harrower in to undertake a health check and colour vision test. Once cleared he was directed to the first examination hall where alongside many other candidates, he spent the next few hours poring over the navigation exam overseen by one of the marine superintendents. The afternoon was spent in another room down the corridor where he was put through his paces answering questions on seamanship, ship handling, emergency procedures, crew handling, ship victualling and a host of other relevant topics. Leaving the offices late in the evening he headed home to await the results.

The results he received when he returned to the Mercantile Office on Tuesday left him stunned. Russell had easily passed all of his exams. Unfortunately, he was informed that his application for a Masters ticket had been refused. When asked why the superintendent informed him that because he had not held a position as a first mate aboard a deep-water vessel for at least 12 months, he was ineligible to receive his Ordinary Master's rating. Russell immediately lodged an appeal against the decision but at a meeting of the Marine Board held on April 6th the superintendents ruling against Harrower's application was upheld. Gutted, angry and bereft of direction, his career aspirations were at a standstill.

Grasping at straws Harrower had tried to appeal to his former captain, James Welburn for a testimonial as to his skills and conduct aboard the Dunstaffnage but this proved to be impossible. Captain Welburn was dead and his ship was a pile of debris scattered for miles along the Scottish coast. The shiny new jute clipper had sat beside the wharf in Dundee harbour for a month discharging her cargo and taking on ballast before Captain Welburn had gained a fresh charter. The Dunstaffnage left Dundee about 8.30 a.m. on March 16th 1883 bound for Liverpool. With most of her voyage sailors discharged she left Dundee with a crew consisting of a master and twenty runners and in tow of the steam tug Recovery. Captain Welburn's wife and two children were also on board headed home.

The report of the loss of the Dunstaffnage gives a concise summary of what happened to the ship to bring about her demise. The clipper had just 608 tons of ballast aboard and was riding high in the water. The royal and topgallant sails had been taken off her but the topsails and courses were loose in their gaskets. She had enough sails bent for the crew of runners to be able to set them in case of an emergency.

"when about 3 miles to the east of the Fair Way Buoy at the entrance to the Tay, both vessels were at about half-past 11 a.m. put upon a N.E. by E. course with the view of going north about, the wind at the time being moderate from the N.W. with a force of about 3 or 4....At

Star of Greece

12.30 it was blowing so hard that they had to prop the tug's wheel at hard-a-port, so as to prevent her head paying off to port. At about 2 a.m. the wind, which had been for some time at N. to N.N.E., suddenly flew round to E.N.E. blowing with hurricane force, and at the same instant the pin, which held the hook and shackle to which the towing line was attached broke, and the two vessels then parted, the hawser with the shackle attached to it going over the tug's stern. At this time we are told that the Girdle Ness Lights bore N.W. by W. distant from 10 to 12 miles... at this moment a blinding snowstorm came on... The snowstorm continued for about an hour and a half... On parting from the tug she paid off, with her head to the N.W., before the gale, and it was observed that her crew were setting the fore topmast staysail, and from that time nothing more was seen of her... between 5 and 6 a.m. the same morning some fishermen of Dunnies, a small fishing village about midway between Aberdeen and Stonehaven, observed a quantity of wreckage along the coast, which, there can be no doubt, must have belonged to the "Dunstaffnage," two buoys and the stern board of one of her boats having been picked up with the vessel's name painted on them.

The wreckage was subsequently strewn along the beach from Findon Ness to Carron Rock, a distance of about six miles, and as the tide began to flow that morning at about 2 a.m., and would carry the wreckage to the south, there is every reason to think that the vessel struck on Findon Ness, and there went to pieces, no wreckage having been found to the northward of that point.... With the wind as it was, from about E.N.E., and the tide setting to the S.W., the vessel would easily do the distance to Findon Ness in a couple of hours, scudding before the gale under her fore topmast staysail, which would bring her ashore by about 4 a.m. The second wave, the assessors tell me, would probably have knocked her all to pieces; and in that case there would be no difficulty in some portion of the wreckage drifting down to Dunnies by between. 5 and 6 a.m. This is, in our opinion the only reasonable solution of the casualty."
Board of Trade Wreck Report for 'Dunstaffnage', 1883, Board of Trade, London.

There were no survivors. The wreck was located half a mile from the shore almost directly opposite where the ship and tug had parted

Star of Greece

company. Caught upon a lee shore in a howling gale the Dunstaffnage drifted onto a reef just a few hundred yards from the beach. Her anchors had been let go but then dragged. Caught in the breakers the iron hull of the ship had been battered to pieces against a notorious reef a little to the north of an island known as May Craig. There was little time for the crew to get out the ship's boats as the decks were swept clean of any not safely in the rigging. It was just a matter of time before the masts went by the board carrying those unfortunate souls aloft into the raging seas. The wreckage, rocks and waves did the rest and all aboard perished in the early hours of the morning dashed to pieces or drowned. A lifeboat and several lifebuoys washed ashore near the village of Portlethen, but no survivors were found by searchers scouring the cliffs and rocky beaches near the scene of the wreck.

Over the next few days, great amounts of wreckage and personal belongings from those aboard began to wash ashore, amongst these were the clothes and toys of a child, and those of the master's wife, Mary Welburn. Two battered bodies later came ashore, one, that of William Grant came ashore barely recognisable, but he was the only one identified, and it was left to his young wife to make a positive identification of the deceased. Those lost were Captain Charles Welburn, first mate James Will, and Seamen; James Elder, carpenter, Martin Launan, C.T. Ostergren, Henry McHunn, John Brightmann, Second Mate, Barney Kelly, William Rollo, boatswain, Robert Bell, Robert Roger, Bates Lamb, Thomas Ferguson, George Hill, John Elder, James Grant, Frank Thomson. Also killed were apprentices William Simonds, Joe Hiildred, John Alexander, and James Welburn, the captain's son. Amongst the dead were Mrs Mary Jane Welburn and her eight-year-old daughter Lily Welburn. The body of Mary Welburn came ashore a week after the wreck and was interred in the cemetery at Fettereso alongside the others from the wreck who had already been recovered.

After attending the inquest into the loss of the Dunstaffnage, where he was not asked to give evidence, Russell Harrower once more returned to the sea. He was in search of a berth aboard a deep-water ship and with experience in the East India jute trade, it was within the ports handling these clippers that Russell carried out his search for employment as a first mate. Turning once again to life aboard a square-rigger Russell found his way to the London Docks. He eventually found himself in the employ of J.P. Corry & Co. By early 1884 Russell Harrower found himself travelling to Hull to take up a position as second officer aboard the Star of Persia, just in from a run to Sydney, Australia. The clipper, advertised as under the command of the long-time Corry captain James Mahood, was to take on a cargo of coking coal, soda ash, soda crystals and whiting. Ill-health prevented Mahood from again taking to sea allowing first officer John McIlroy was to take over. This was the opportunity for Harrower to gain the much-needed square-rigger sea time he needed. The Star of Persia stayed at Hull for more than a month before being cleared out on April 29th 1884 from the William Wright Dock and towed to the departure moorings of the outer Humber.

Departing Grimsby on May 1st the Star of Persia sailed down the coast towards Deal pushed along by light breezes from the north-northwest. Conditions were hazy and overcast and Captain McIlroy instructed his lookouts to keep a sharp eye forward in the crowded sea lanes. The winds backed around to the southwest and freshened as a cold front moved in from the west. The squall lines convinced McIlroy to seek temporary shelter in the Downs until the worst of the weather passed. The squally winds and showers forced the Star of Persia to stay put until the wind shifted again this time to the northeast. They departed the Downs on May 4th whilst the wind held, predicting that it would shift again as the barometer rose steadily. He was correct in his assumption as the breeze came fresh and cold from the northwest. The clipper had barely cleared the Chops of the Channel before yet another change wandered through from the northwest. The cold front brought with it the usual dirty weather and seas as the Star of Persia thrashed her way across the Bay of Biscay and into deeper waters.

McIlroy was attempting a westing run around Cape Horn in the middle of a southern winter. The Star of Persia ran into a series of violent northwest gales as she ran down the coast of Patagonia. The clipper shipped a great deal of water over her bulwarks with the usual damage on deck with sails shredded and crew washed into the scuppers. The weather and sea conditions only worsened as the vessel attempted her westward run around the Horn. A series of Cape Horn

Star of Greece

snorters greeted those aboard the Star of Persia along with the mountainous greybeards that threatened to drive the struggling clipper below the grey-black seas. The vessel battled onwards for more than a week taking a fearful dusting as Captain McIlroy and his crew fought the horrendous weather and seas. Running well below 50° south the Star of Persia sailed almost dead south-southwest through Drake Passage before a favourable slant was found upon the tail end of an intense low-pressure cell. The ship cracked on northeast as she rounded the Horn before sailing up the west coast of South America.

The winds, at last, hauled around to the west freshening as a cold front pushed the clipper rapidly westwards past the Farallon's towards her final destination. The Star of Persia hove to in the soundings near the 'potato patch' located on Four Mile Bank a few miles outside the Golden Gate entrance to San Francisco Bay. There the ship adrift, her yards backed, awaited the San Francisco Bar Pilot schooner to appear so she could be guided through the Heads and into the sheltered waters of the Bay. The clipper was safe across the bar on September 23rd, 142 days from Deal, and was hauled into the anchorage to await her discharge orders. Even before the ship had rounded up to drop anchor the notorious crimp-boats were racing out to greet the clipper. Shipping men at £2 10s a month it was little wonder that many of the crew jumped ship and into the hands of the boarding house runners in hope of better pay aboard an outbound vessel. The mooring of the Star of Persia was left to those few loyal sailors, the officers and apprentices who remained. The clipper was hauled into Oakland Bay coal wharf to discharge. The harbour was filled with ships awaiting a return cargo and it was three weeks before the Star of Persia was hauled into Oakland's Howard Street Number 2 wharf to take on her outward cargo.

California's cereal crops drew hundreds of British vessels to San Francisco every year. The 14000 miles back to Britain meant little to the dried grain deep in the holds of the sailing ships headed home for orders. The iron and steel clippers acted as floating grain bunkers and the cargo drew a premium price in ports across Europe. The clipper was towed up the northern end of San Francisco Bay to Benicia on October 23rd to complete loading. The crew eventually battened

Star of Greece

down hatches on 1400 tons of bagged wheat worth nearly $48000 US, and 348000 lbs of copra being transhipped from the Pacific. The ship was cleared into the stream from Long Bridge on November 10th 1884 and set sail through the Golden Gate on November 15th bound to Falmouth for orders. The run home was easier through the southern latitudes and around the Horn. Falmouth was reached on March 26th 1885, 132 days out from San Francisco, where McIlroy received orders to immediately sail for Dublin to discharge.

The Star of Persia put out from Falmouth on the morning of March 27th, Captain McIlroy steering southeast as freshening south to southwest winds blew up the Channel. When far enough out the clipper turned west clearing the Lizard later that afternoon. The fresh to strong south-westerlies brought with them periods of rain and very rough head seas that made sailing difficult in the approaches to St George's Channel. The wind shifted round to the north as the Star of Persia loped up towards Dublin forcing John McIlroy to tack his ship through the narrow sea lanes. The outer soundings of Dublin Bay were reached on April 1st as the clipper hove to awaiting a pilot to guide her in past the treacherous sandbanks and shoals of the outer harbour. Taken in by steam tug the Star of Persia was soon warped into the Northern Wall.

Orders arrived for the Star of Persia to load coal from Wales, and the clipper was towed from Dublin Bay on May 16th and thence up the Bristol Channel to Cardiff where the Star of Persia dropped anchor on the 19th. There the tug was directed to pull the lightly ballasted ship into Penarth coaling wharf to take on 1800 tons of coke and soda for Mauritius. Taking her place beneath the coaling chutes the Star of Persia was rapidly loaded as the crew made final preparations departure. Harrower found himself promoted to first officer, with the third mate, 20-year-old John Hazeland taking up Russell's former position.

Captain McIlroy was still in command as his ship was towed back down Bristol Channel on June 9th 1885 outward bound for Mauritius. The Star of Persia set sail on a light north-westerly breeze and smooth seas. The clipper enjoyed fine weather for much of the way to the Cape of Good Hope and the winter westerlies made for excellent if rather rough sailing conditions as the ship rounded the Cape on her run towards Mauritius. The southeast trades came on strong and steady, the Star of Persia reaching the outer roadstead of Port Louis on August 20th, 72 days from Cardiff. The ship was soon hauled into the inner wharf to discharge.

After a quick sprint across to Saugor, the clipper was just three weeks to Calcutta thanks to the steady monsoon winds. It was on this voyage that Russell Harrower finally worked up enough sea time as a first officer to meet the minimum requirements of the Board of Trade's regulations for his Ordinary Ship's Master Certificate. He telegrammed his brother Frank in London instructing him to take his qualification papers to Corry's office who would then forward the paperwork along with a copy of the Star of Persia's brief log to the BoT for verification.

It took some time for the application to be processed but at last more than two years after he had sat for and passed the Board of Trade's Ship Master's examinations, Russell Harrower was issued his ticket on December 19th 1885. He was not at home to receive it, still at sea as Mate of the Star of Persia. There was perhaps a genuine sense of pride in the Harrower household at Dartmouth Park Hill. His parents had separated and now Jemima Harrower lived in Highgate with her son Frank and his family. Russell would stay there whenever he was home from the sea and his promotion would mean a greater income, perhaps £50 to £100 in the kitty every time he set out as master of a deep-water trader. At 27 Russell Harrower was quite young for one of Corry's potential captains and his current job was the equivalent of being on probation, a chance to prove his credentials to the owners and managers in Belfast and London. He simply had to wait for a berth to become available, a frequent occurrence as more and more of their veteran skippers either retired or moved into steam-powered vessels.

The Star of Persia arrived at Gravesend from Calcutta on March 24th 1886 her Master, John McIlroy deciding to come ashore after years at sea. The ship was towed into Millwall and warped into her berth to discharge. With his ship now safely moored Captain McIlroy stepped ashore and for a time said goodbye to a life at sea.

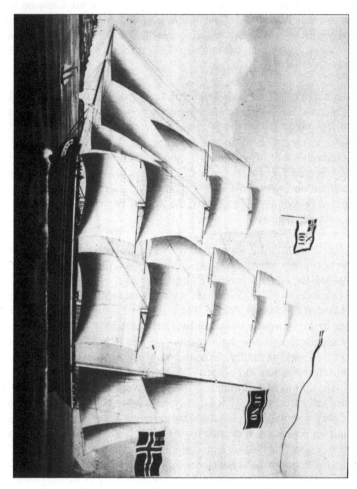

The Juno under the Norwegian flag.
Unknown artist.

Star of Greece

In his place, came Captain Henry Russell Harrower, newly promoted from first officer. The young ships' Master took his new command into the East India dry-dock for a much needed clean and refit. The Star of Persia was in dock for almost a week, Captain Harrower eager to make sure that nothing was left to chance when he took the clipper back to sea. With several hundred tons of ballast aboard the vessel was towed from Millwall on April 21st, down the Thames to Cardiff to take on coal for Mauritius and Calcutta.

Another of Corry's ship, the Star of Denmark was outward bound to Adelaide, South Australia filled with general cargo for the colony. As the sun set on April 9th, 79 days from St Catherine's, a fearful glow was spotted upon the horizon signalling to all aboard that most feared of mariner's horrors, a burning ship at sea. The blazing vessel was reckoned by Captain Thomas G Steer and his officers to be no more than 40 miles ahead and the orders were given to make all sail. Both watches were sent aloft and as the barque approached the burning wreck Captain Steer ordered the boats hauled out onto the davits and made ready for lowering in case they had to be used to take off any survivors. As morning approached the Star of Denmark sailed to windward of the blazing hulk, at position latitude 36°33' south, longitude 11°24' east.

The dawn brought with it a truly horrifying sight for a sailor, a ship ablaze from stem to stern. A thunderous roar could be heard as enormous flames and billowing clouds of thick smoke reached skyward. Tom Steer tacked about the ship to conduct a complete survey of the stricken vessel. Sadly, no identifying marks could be observed such was the damage already done. The only spars standing were the stumps of the fore and mainmasts. The mizzen mast and rudder head were already destroyed. One bower anchor was hanging off the fo'c'sle head whilst the other was still stowed. The ship appeared to be of a composite construction of about 1200 tons, her copper sheeting, buckled and peeling away above the waterline. Captain Steer kept lookouts aloft and had his vessel standing by the blackened hulk for ten hours on the off chance that any boats that had put off from the wreck were still in the vicinity. After an extensive delay with no survivors in view, the Star of Denmark took one more

Star of Greece

turn about the doomed vessel before again making sail and rolling away southwards.

Unbeknownst to those aboard the Star Liner the stricken vessel in question was the Swedish ship Juno, 1248 tons, outbound from Bergen to Melbourne under the command of Captain T. Keyller. The Juno had caught fire on April 8th at 34°44' south, 11°34 east and was soon after abandoned by her crew who set out for the west coast of Africa. The ship's boats were at sea for nine gruelling days before they made landfall 40 miles north of Port Nolloth, a rudimentary copper port on South Africa's northwest coast. As the boats attempted to make the beach they were caught in the treacherous surf and capsized. Those aboard were dumped into the boiling waters of Alexander Bay, and of the 22 sailors who had successfully evacuated the burning ship Juno, only four made it to shore alive. The exhausted survivors eventually staggered into Port Nolloth and soon after were taken by the 182-ton steamer Namaqua to Cape Town.

The Star of Persia arrived at her Cardiff berth on April 23rd and took her place in the line of vessels waiting to load at the Penarth coaling staithes. The ship was cleared out by May 2nd to the departure moorings of Ely Tidal Harbour. A tug took the heavily laden vessel out past the breakwater and into the roadstead on May 3rd where Harrower welcomed the pilot and the tug that would tow the clipper down the Bristol Channel. The wind was freshening from the south and southeast with showers and periods of rain punctuating the start of the voyage. The tow line was let go off Lundy Island as Captain Harrower made his course east by southeast to clear the Scillies. The Star of Persia headed for open water as the wind turned southerly bringing with it squalls and high seas. As the clipper continued southeast a dead muzzler forced Captain Harrower to tack again and again across the Bay of Biscay.

Clearing Spain's Cabo Villano and the reefs of the Galicia Bank Russell Harrower gave orders for all plain sail to be set hoping that the northeast trades would soon be picked up. The winds, when they did put in an appearance, proved light and unsteady to the line which was not crossed until 30 days out. The southeast trades were a tad more reliable and lasted until the 18th parallel when they fell away

Star of Greece

altogether. A continuous easterly wind blew steadily for several days allowing the Star of Persia to make up time as Russell Harrower cracked on south towards the Prime Meridian. Crossing out of the easterlies a series of violent south-easterly gales were encountered as the clipper sailed south by east headed towards the Southern Ocean. These were soon replaced by prevailing winter westerlies as the Star of Persia began her easting run. As the gales passed Captain Harrower took advantage of clearing weather to steer northeast towards the Indian Ocean. The ship sailed straight onto Port Louis the southeast trades coming in steadily off the starboard quarter.

Arriving in Grand River Bay the Star of Persia was shifted to the pier and stevedores quickly set to discharging her much needed cargo of coal. It took almost a fortnight to empty and clean the holds, then take on enough ballast and sugar for the trip to Calcutta. The southwest monsoon provided the vessel with a rapid if rain and squall filled passage to Saugor where the pilot brig greeted them with the approaching cutter's usual alacrity. Two days tow up the Hooghly saw the Star of Persia again at her regular berth. This time she was not there to load only jute, those days were all but over. Instead, Captain Harrower was under contract to take on grain, sacking, cotton cloth, tea, preserved foods, jute, linseed, and exotic trade goods from the interior. Jute was being shipped at just 30 shillings per ton and was proving to be less than profitable as Corry's ships plied the colonial trade routes more and more. Russell Harrower kept the Star of Persia in port just long enough to load for home. The ship left Calcutta at the beginning of October, her run to London almost completely free of incident. The pilot schooner was met off St Catherine's on December 21st and a tow arranged soon after, making for a round trip of 7 months and 17 days. Not a record run but well within the average for the Star of Persia in the halcyon days of the Calcutta jute trade.

Captain Harrower's troubles only really began as the ship was headed inbound to the Downs anchorage with crew aloft to clew up the sails. The pilot ordered the helmsman to take the clipper astern of two other vessels already at anchor. With the crew making preparations to let the anchors go someone had forgotten to light the port and starboard navigation lanterns. The hands on the fo'c'sle head were distracted

147

Star of Greece

by their roles in getting the anchor and 45 fathoms of cable on a spring ready to be let go. Ahead at anchor was the 738 ton, Italian barque, Draguette, with an anchor-light hoisted above on the forestay. The weather was rainy but with good visibility to the horizon. Winds were blowing fresh from the south-southwest and the tide was running northeast at 3-4 knots. The Draguette, on her way from Amsterdam to Java with general cargo, was anchored between Deal Pier and Walmer, 1¼ miles from shore, starboard anchor out and her head to the wind.

The Star of Persia was observed by the lookout aboard the Italian barque at no more than a ship's length away approaching at about 5 knots from the south, no navigation lights showing. She was seen to be headed towards the shore almost right ahead of the Draguette, sailing athwart the tide. The mate of the Draguette attempted to warn off the approaching ship, hailing her helmsman and madly ringing the ship's bell to get the Star of Persia to veer off. Alarmingly the efforts of the Draguette's crew were to no avail as the Star of Persia bore down upon the helpless barque. Too late Captain Harrower ordered the helm hard over, the bow coming around just as the two vessels struck.

The crews of both vessels cried out in panic as the Star of Persia's starboard anchor became entangled in the Draguette's fore rigging. The whole assembly was torn clear away by the anchor which took with it the barque's bowsprit at the knightheads. The ship then swung sharply around on her cutwater striking the Italian barque's stem and port bow, stoving in the latter's plates and bulwarks as she passed. The Star of Persia's anchor was ripped out of the stoppers and the topgallant bulwarks abaft the break of the fo'c'sle were carried away. Both vessels let their chains run out, the crew aboard the Star of Persia being forced to slip the cable leaving their bower anchor entangled in the Draguette's mangled fore-rig. With her forestays carried away the Draguette's foretopgallant mast was in very real danger of coming crashing down. The Star of Persia left a crippled victim in her wake. Russel Harrower had his vessel quickly hove-to her, sails backed as the crew worked to get the remaining anchor rove to the cable and catted. The anchor was soon let go and Harrower and his officers set about surveying the damage done. The carpenter sounded the well but the ship appeared to be taking no water. Both vessels stayed at anchor overnight and into the next day as the winds freshened eliminating a tow into the Thames. The crippled vessels were towed into port late on the 23rd, and as the Draguette was under a salvage claim her owner Captain Schiaffino of Genoa was in a truly dark and vengeful mood.

The Star of Persia limped into the East India Dock to discharge, her owner's none too happy about the damage their newest skipper had caused to the ship. The clipper was quickly unloaded and taken into the dry dock for repairs and a refit. Captain Harrower was not the only Corry skipper to be in the bad books that Christmas. The Jane Porter under 38-year-old Jerseyman, George du Gruchy, with a load of wheat from Portland Oregon, collided with the 560-ton, Newcastle steamer Spring Hill as she came to anchor at Queenstown. Both vessels suffered extensive damage and had to be towed into port for discharge. Each collision required adjudication in the courts, Captains Harrower and De Gruchy both having to appear in multiple cases involving insurance and salvage claims by the owners of the damaged vessels. Captain Henry Russell Harrower was called before the High Court, Admiralty Division to settle the dispute between the owners of the Draguette and Star of Persia. Appearing before Justice Butt and the Trinity Masters the two sides put their cases and after several hours of testimony, with each side blaming the other for the collision, Justice Butt delivered his findings.

"the conclusion I come too, so far as one can judge from the demeanour of the witnesses, is that the story told by the witnesses of the Italian barque is a credible story. On the whole, I come to the conclusion, without much doubt or hesitation that the case of the plaintiffs is established, and I must announce the Star of Persia is alone to blame."
Justice Charles Parker Butt – High Court Judge.
New Zealand Herald, Volume XXIV, Issue 7901, 21 March 1887, Page 4.

Captain Harrower was not yet off the hook as he was called to appear at the Salvage Court the very next day as the tug captain who had taken the Star of Persia into Gravesend had put a claim to rights over the recovery of the clipper and her cargo. Appearing again before Justice Butt, Corry's lawyers managed to avoid costly salvage fees by successfully arguing that the Star of Persia had in no way had her means of propulsion impeded and was in no danger of foundering. Justice Butt did find that the tug had provided timely assistance to the Star of Persia and thus awarded the tug owners, £150 salvage fees plus costs. It was a small victory for Captain Harrower who otherwise may well have lost his job. As it was Justice Butt censured Captain Henry Harrower and ordered him to pay the costs. The Star of Persia was soon repaired and placed on the berth to load for Sydney. Captain De Gruchy was also found at fault at the enquiry held in Queenstown in the Case of Jane Porter vs Spring Hill. The two disgraced skippers faced further reprimand from Corry's agents in London. Russell Harrower lost command of the Star of Persia and was placed in charge of the newly repaired barque Jane Porter. George De Gruchy was forgiven his transgression and sent down to Millwall to take over the Star of Persia for her run out to Australia. After being repaired and refitted at the yards of Harland & Wolff in Belfast, the Jane Porter was towed across to Liverpool to take on coal for Calcutta.

Russell Harrower saw in the New Year at home in High Gate before taking the train to Liverpool in mid-January to oversee the loading of his latest command. In a backhanded manner being made master of the Jane Porter was a show of ongoing confidence by J.P. Corry & Co' as almost all of their best captains had been in charge of the little barque before being given command of one of Corry's crack clippers. The Jane Porter was in fine condition as Harrower arrived to inspect the vessel. The ship's holds were almost filled when Russell stepped aboard at the end of January 1887. Final preparations were soon made and the Jane Porter was warped out into the stream to be towed down the Mersey to the departure buoys on February 11th. The clipper was up and away in the morning towed down the river upon the rising tide. The steam tug took the barque out into the Irish Sea and left the Jane Porter just north of Holyhead on the 13th.

The vessel sailed south into St Georges Channel as an intense low-pressure system moved in across from the west bringing snow and sleet showers. Seas came on rough to very rough as the barque crashed her way south under topsails and headsails.

Victoria Terminus, Bombay, circa 1879.
Clifton & Co', Bombay.

Star of Greece

Winds moderated and swung around to the northeast giving the Jane Porter a favourable slant as she passed southwest of the Lizard and into the North Atlantic. Freshening nor'easters pushed the clipper rapidly south and she made the island of Madeira in double-quick time.

The trades stayed fresh and strong to the equator, which was crossed in mid-March. Once through the doldrums, the southeast trades lasted until 30° south. Captain Harrower took his ship well out into the South Atlantic crossing below the Cape of Good Hope along the line of 43° south before shifting course north by northwest to take the Jane Porter up into the southeast trades. Crossing the southern Indian Ocean through late April the winds proved light and variable making for slow progress up through the ocean's middle. Entering the Bay of Bengal there was little wind to speak of as the barque ghosted across the Bay. The voyage was without major incident and the Sand Heads pilot station was at last sighted on May 31st, 107 days from Holyhead. Arriving at Calcutta on June 2nd the Jane Porter had barely been snugged down to await her discharge berth when Russell Harrower received a telegram instructing him to make all haste back to London to take over command of Corry's flagship, the Star of Greece.

Harrower was forced to make hurried arrangements for the handover of the Jane Porter to her new captain, A.A. Walker, before stepping ashore and collecting his ticket home from Corry's Calcutta agents. He headed over to the booking agents for P&O (Peninsular & Oriental Steam Navigation Co.) at Garden Reach to check the train timetable. There was an express mail train leaving from Calcutta in a few days that would connect with the P&O steamship Assam, in Bombay. The timing would be tight but Russell Harrower had little choice if he wanted to make London before the end of July when the Star of Greece was scheduled to depart. The train steamed out of Calcutta exactly on time and took Russell through the hot and dusty interior of India taking him to Allahabad where he changed trains to journey down to Jubbulpore in Madhya Pradesh.

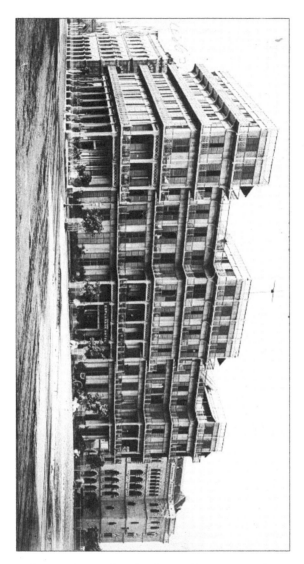

Watson's Hotel, Bombay in the 19th century, circa 1890s.
Unknown photographer. Old Indian Photos.

Star of Greece

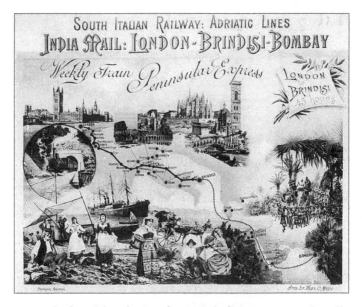

Postcard advertising the London to Brindisi express service. Circa 1880's.

P&O Royal Mail Peninsular-Express in Italy, running from Brindisi to Calais, circa 1890.
Collection of .Dr. Bruno Bonazzelli.

Star of Greece

The city was the meeting point for the Great Indian Peninsula Railway, a distance of 616 miles from Bombay, and the East Indian Railway, 220 miles from Allahabad, and formed a major hub for the Bombay-Calcutta railway line. At Jubbulpore Captain Harrower spent the night at a Railway Hotel before boarding a GIPR steam train for Bombay. It was to be another two days before the tired traveller reached his destination, the Victoria Terminus in downtown Bombay. Alighting from the train Captain Harrower was met by another of Corry's agents who accompanied by a series of immaculately dressed railway porters escorted Russell across the square to Watson's Hotel. The five-storied cast-iron edifice contained over one hundred guest rooms, a luxuriously appointed lobby, fine dining restaurant and an exclusive bar at the ground level. Here the weary traveller stayed until the steamer Assam was ready for her departure. It had taken just five days to travel from Calcutta and the rapidity of the journey was in stark contrast to the long and hard days spent taking a jute clipper across the globe.

P&O's steamer Assam, under Captain W.A. Wheeler, was loaded and ready to depart Bombay on June 7th 1887. The monsoon was in full swing and it looked like thick and dirty weather was in the offing as the ss Assam sailed around the Colabah Peninsula and onwards to the coaling port of Aden. The weather worsened the further out into the Arabian Sea the Assam sailed. An intense tropical low saw the ship battling cyclonic winds and huge angry seas in the approaches to the Gulf of Aden. At several points during the worst fury of the tempest, the Assam shipped huge amounts of water over her bulwarks which flooded the decks causing tremendous damage. The galley doors were stove in and the kitchen gutted of all its contents including the Goanese cooks. The skylight above the saloon was also carried away causing flooding of the dining room and adjoining cabins below. Conditions worsened as bulwarks were carried away and chicken coops, the pigsty and cows in the pens were washed overboard.

Just to add insult to injury the bakery was washed out and cold meals had to be served to guests and crew alike whilst the storm raged. The crew of Goanese stewards and sailors aboard the Assam worked tirelessly to keep the passengers safe and well-fed. Despite the very real injury to her topsides the Assam resolutely steamed on arriving safely at the port of Aden on June 16th. The ship stayed just long enough to top up her coal bunkers before again setting forth up through the balmy climes and calm waters of the Red Sea. The Port of Suez was arrived at on June 21st and the Assam was forced to await her scheduled run up through the Suez Canal and onto Port Said. From Port Said, the steamer sailed onto Alexandria to take up the last of her mail, passengers and coal. She did not depart Alexandria, Egypt until Tuesday, June 28th at 6:30 pm, bound for Brindisi, Italy, having spent several days in port having her damaged fittings and cabins repaired.

The P&O liner SS Assam (ex- Feldmarschall Moltke).
Rabson & O'Donoghue, World Ship Society Collection.

The Assam arrived at her Brindisi P&O berth on July 1st at 1:30 pm and the Brindisi to Calais Express Mail Train left the same day at 5:30 pm. The trip to Calais took 44 hours and upon arrival, Russell Harrower wasted no time in gathering his belongings and catching a ferry bound for Dover. Even as he was crossing the English Channel the Star of Greece lay at anchor in the Downs awaiting her new captain. With the delayed arrival of her new captain confirmed by telegram from Alexandria, the first officer, John Hazeland, of the Star of Greece had the clipper hauled out from her Millwall berth and towed down to Gravesend on June 28th. A pilot and tug were arranged and the ship was brought to the Downs the following morning. Russell Harrower was still at sea aboard the Assam and it would be another 3-4 days before he would arrive at Calais. Hazeland was hoping that he would be given command of the ship in the absence of her newly appointed master.

Star of Greece

Clear skies and smooth seas saw an almost empty anchorage as those aboard the Star of Greece awaited the prodigal captain's appearance. As fine and warm weather settled in over the south of England the decision was made to move the Star of Greece from Goodwin Sands to East Wear Bay anchorage to await Captain Harrower's arrival. The Brindisi to Calais mail train arrived on the afternoon of July 3rd and Russell Harrower took the first ferry to Dover where he was greeted by Corry's agents who arranged for a pilot steamer to take him and his dunnage out to the waiting ship. Once aboard Harrower took over immediate command, bruskly thanking his obviously disappointed first mate. The Star of Greece took up her tow early that evening as the breeze blew from the west heralding an approaching change.

Star of Greece

An Unusual Cargo!

Even as Russell Harrower was taking his new command out to Calcutta another Corry ship was ending her career in a hauntingly tragic and prescient manner. The Star of Scotia, commanded by Captain Michael Coulter with a crew of twenty-one hands, was on a voyage from San Francisco to Queenstown for orders. Aboard she carried 33784 centals of wheat valued at $56433 US destined for the European markets. The 1000-ton iron clipper was running before a blustery gale sailing north by east as Captain Coulter attempted to clear the Falkland Islands. It was about 11:30 pm on the night of June 28th 1887 when the clipper ran aground on Bull Point, 100 miles east of Port Stanley, in relatively fine weather. A Northern Ireland paper, the Belfast News-Letter provided a graphic description of the wreck in heart-rending detail;

"It was half-past eleven at night when the vessel struck and the crew remained on board until daylight. The weather was somewhat calm but a heavy surf was breaking on the island. The captain, with the majority of the crew, got away in one boat, and the second boat was in charge of Mr Frazer. The captain told the men not to bring their effects but to save their lives and then go off on the ship again for their clothes. The captain's boat was the first to reach the shore, but the mate's boat was cast upon the beach under heavy surf, and in it were the semi-conscious forms of two of the crew – an Englishman and a Scotchman named Davis and Drummond. The mate was also in the boat, but he had just breathed his last.

Davis and Drummond both recovered and gave an account of their shocking sufferings during the short time that had elapsed since they had abandoned their vessel...while they were making their way along the shore the boat capsized and all of the occupants thrown into the water. Some of the men managed to cling to the upturned craft which afterwards righted. Some of these poor fellows never reached her again. While making once more for the shore the boat was again capsized – and this happened no less than seven times until there were only three left – the Mate (Mr Frazer) Davis and Drummond. Some

Star of Greece

of the unfortunate men had during the time clung to the boat but their strength had failed them. Then becoming exhausted they had to loosen their hold and perished in the water.

The water...was intensely cold, and it was only by their robust condition that....Davis and Drummond saved their lives. They saw their comrades drowning...Where the shipwrecked men landed the island was covered with snow, and the place altogether bitterly cold. They found shelter in the hut of some herdsman...The herdsman rode 100 miles to inform authorities...the bodies of four of their comrades were washed ashore and later on the bodies of two others were found on the beach...and the bodies presented a shocking spectacle...The birds had come across the bodies of the two shipwrecked men and were short but complete in their voracious work...the night following the stranding a heavy gale sprang up and in the morning the Star of Scotia had disappeared...The men...had to remain for a month on the island before they were taken away by a German steamer. "

Belfast News-Letter *05 September 1887*

The final death toll from the Star of Scotia wreck was seven men; first mate William Fraser, 21 of Newcastle, Apprentice Arthur Burton, 17 of Chelmsford and five Able Seamen, John Anderson, 30 of Sweden, Edward Casey, 27 of Limerick, David Parkinson, 31, and David Roberts, 25, both from Jamaica, James Turner, 44, of Glasgow and Guss Willis a 22-year-old sailor from Georgia in the United States.

Winds held from the nor 'northwest as the Star of Greece romped down the Channel having taken her start off St Catherine's on the morning of the 4th of July 1887. She had fine sailings winds punctuated by lines of showers and thunder-squalls all the way 'to the Chops' when the wind backed around to the west and freshened considerably. Passing north of Ushant on the evening of the 6th the Star of Greece sailed into a dead muzzler and Captain Harrower was forced to tack back and forth into a rising sea as his ship crossed the outer reaches of the Bay of Biscay. The winds stayed fresh to strong from the northwest for several more days until brisk northeast trades gradually took over allowing Captain Harrower to stack on the canvas and push his ship to its limits. The perfect sailing conditions

Star of Greece

continued for more than a fortnight, Cape Verde being passed on July 21st just 17 days from St Catherine's. The trades petered out just north of the equator to be replaced by variable winds and thunder squalls. These lasted until the southeast trades were picked up on July 29th. The winds proved fresh and strong allowing for some fine sailing as the Star of Greece charged on out into the South Atlantic. Passing below Tristan da Cunha the westerlies came on with frustrating variability as Captain Harrower drove his ship south by east into the higher latitudes.

The meridian of the Cape of Good Hope was passed on the 18th of August as the ship made her easting run between 40° and 43° south. There were periods of boisterous weather as the clipper rolled on the northeast by east into the Indian Ocean picking up the southeast trades as she crossed the 20th parallel. The trades stayed strong and steady all the way north as the vessel raced towards Ceylon. The monsoon winds were blowing from the southwest as the Star of Greece cruised on past the island. The run up into the Bay of Bengal was punctuated by squally showers and thunderstorms as the clipper raced away north by northeast. The shallow waters of the southern Long Sands were reached on September 25th.

It was the height of the cyclone season but fortunately for those aboard the Star of Greece she had managed to avoid what had been a particularly nasty period of intense tropical storms. It was whilst she was berthed taking on 1800 or so tons of linseed, rice and wheat that her crew witnessed a rather unusual spectacle. The iron, four-masted barque Palgrave, the largest sailing vessel of her type in the world at her launch, was anchored in the Hooghly having come into Garden Reach on September 16th. Weighing nearly 3200 tons, carrying a deadweight of 5000 tons, and being over 322 feet long, her tugs had a difficult time bringing the enormous barque to berth at Prinsep Ghat. There she was to load jute and grain and was one of the first of the new breed of slab-sided windjammers that were rapidly replacing the finer lined clippers of the 1860s and '70s. Between these new iron and steel behemoths and the growing reliability and economy of large steamers, the age of the iron clipper was coming to a rather ignoble and economically rational end.

Star of Greece

The Star of Greece was loaded up with grain and ready to depart Calcutta by October 30th, clearing out on November 1st, 1887. She was towed out from her Garden Reach berth the next morning arriving in the Saugor roads on the 4th. The monsoon winds were in their transitional phase with the southwest winds gradually retreating to be replaced by coastal northerlies. Captain Harrower decided to take advantage of these fickle winds and sailed the clipper south by west down the Coromandel Coast towards Ceylon passing Cape Calimere five days out from Sand Heads.

The ship enjoyed fair weather to Cape Agulhas. Fresh to strong easterlies allowed Captain Harrower to gain a favourable slant around the coast of Africa and up into the South Atlantic. The summertime airs of the southeast trades proved strong and steady as the Star of Greece raced past St Helena and onto Cabo Sao Roque by the end of December. The ship lay becalmed for several days north of the equator until the northeast trades were picked up. These proved light and variable as Captain Harrower steered a course north by west towards the Cape Verde Islands. The weather worsened rapidly the further north the clipper sailed. A series of northerly gales created atrocious sailing conditions as the Star of Greece approached the Western Isles.

Harrower was forced to sail well out to sea to avoid being caught on a lee shore. Very cold and dirty weather accompanied by snow showers battered the Star of Greece as she thrashed her way homewards. It was at the height of one of these storms that the ship nearly came undone as she rolled home in heaving seas. The clipper was shipping huge amounts of water across the deck, living up to her reputation as a ship fit only for seagulls. The crew were struggling to keep the clipper from being swamped or broaching to when a violent squall caught the Star of Greece flat-aback. There was a thunderous crack as the foresails shredded followed quickly by the tearing and grinding of iron as the foretopmast carried away. The toppling mast brought with it the upper topsail, topgallant and royal yards.

Star of Greece

**Four Masted Iron barque PALGRAVE, 3174 tons, built 1884 by
William Hamilton & Co, Port Glasgow.**
Brodie Collection, State Library of Victoria.

Star of Greece

Of course, as the mast fell it took down the main topgallant mast along with royal and skysail yards. The tangled mess crashed to the deck as both watches were called on to clear up and cut away the wreckage. Axes, mauls and cold chisels were quickly distributed as Captain Harrower ordered the ship hove to. The crew worked furiously to clear away the fallen rigging before the iron lower yards punched a hole through the side of the hull. The Star of Greece rolled dramatically in the heaving North Atlantic swells as the ship was gradually set to rights. It was some days before a spare royal mast and yards could be sent up and a jury foretopmast stepped in and fished to the stump at the doublings. To this Captain Harrower had attached the recovered topsail and lower topgallant yards to which were bent the spare sails taken up from the lazarette. Both watches worked masterfully to bring the ship back into shape. Before the week was out she squared away north for the English Channel.

The embattled vessel came into the soundings from the west rolling along under shortened sail with a quartering nor' westerly breeze. The Star of Greece hove-to off of St Catherine's to take on a pilot amid a raging snowstorm, on February 11th, 99 days from Saugor. The clipper rounded up into the Downs to await a tug. Gravesend was reached towards midnight and the anchor was let go to await the morning tide. With her damaged rigging and jury foretopmast, the Star of Greece looked a right shambles as she was taken up to her berth in the East India Dock on February 13th. With the first mate's words of "That will do men!" the voyage was done.

With the Star of Greece safely docked, Captain Harrower left the unloading of his ship to the care of John Hazeland. Upon reporting to Corry's offices in Fenchurch Street he learned that the Star of Greece was to undertake a special charter and would require a custom fit-out before being placed on the berth to Adelaide, South Australia. The announcement of the voyage to Adelaide was welcome news, Russell would finally have the chance to be reunited with his sister Alice, her husband James and their growing brood. Upon the discharge of her cargo of grain, the Star of Greece was taken into dry dock for repairs to her rigging, a general clean and alterations to her main lower hold.

The top image is of an Armstrong BL 9.2 inch Mark VI breech-loading gun. The bottom image shows the elevating mechanism for the gun. Both are the same as those ordered by the South Australian Colonial Government for the never-built fort at Glenelg.
Nepean Historical Society Collection.

Star of Greece

These involved the setting up of iron mounts and bracings designed specifically to hold steady a wooden crate containing a 24-ton long gun and its mountings, destined for a fort overlooking the Adelaide foreshore. Once at home Russell was greeted warmly by mother Jemima, Frank and his family. In the days leading up to the Star of Greece's departure Jemima Harrower prepared letters and packages for her daughter and grandchildren, some she had never met. A telegram was sent ahead to the Bishop household in Woodville, Adelaide, announcing Russell's impending arrival in the colony.

Soon the refitted the Star of Greece was shifted around to the South West India Dock on March 1st 1888 where an oversized barge awaited her arrival. Upon the craft, packed in a sturdy wooden crate was one of two shore-based guns destined for an as yet to be built fort at Glenelg. The gun was a made to order BL 9.2 inch Mark VI breech-loading cannon that fired a 380-pound projectile with an effective range of 9000 yards. Set upon a special hinged mount, it was known colloquially as a disappearing gun. The gun, manufactured by Armstrong, Mitchell & Co', would be mounted in an emplacement that provided overhead cover which allowed the gun to be loaded when it was below ground level. This design provided maximum protection to the soldiers operating the gun. As the cannon was below ground level a warship approaching the Glenelg Fort would become aware of the weapon only as it was raised and fired.

The cost of shipping the 24-ton monster from London to Adelaide came to an astonishing £400 due to the unusual nature of the charter and its inherent insurance risks. Jute of the same weight would only cost around £36 to ship. As the barge was warped in alongside the Star of Greece special heavy tackle was attached to the oversized crate from the lower yards of the clipper to prevent it from swaying about and damaging the ship. An enormous crane hoisted the gun from the barge and with the help of local stevedores, it was carefully lowered through the main hatch and into the custom-made cradle deep in the lower hold. The total weight of the gun and case came to more than 25 tons, and along with the 6 tons of mountings and casings provided a rather novel site to those ashore. The awkward nature of the cargo made the ballasting and trimming of the clipper difficult as she then took on the rest of her general cargo. If the cannon broke loose during

Star of Greece

rough weather there was a very real possibility that the 24-ton behemoth could punch a hole through the iron plating of the ship, sending and her unsuspecting crew straight to the bottom.

The loading of the Star of Greece was completed by March 15th and she was soon cleared out ready for sailing. The signing on of a crew was also proving to be something of an issue as London shippers were offering just £2 to £3 a month plus a month's advance that would have to be worked off, a situation known as 'flogging the dead horse' before they began earning a real wage. Even as Harrower was making final preparations aboard ship, the local boarding house runners were rounding up the last of the crew needed to make up the ship's complement. In the end, the Star of Greece departed Millwall with a crew of 27 onboard. This group of men comprised Captain Harrower, John Hazeland - First Mate, William Waugh - Second Mate, Charles Commerford - Third Mate, William Parker - Boatswain, the idlers Robert Donald – Carpenter, Gustaf Carlson - Sail Maker, George Blackman – Steward, and Blackman's young protégé and apprentice, George Carder – Cabin Boy and Cook. There were six apprentices berthed in the halfdeck house; Frank Kearney, 19, from County Cork, Edward McBarnett, 18, from Wellington, New Zealand, Alfred Prior, 17, from Norfolk, James Johnstone, 17, from Locharbie, Fred Skeggs, 18 from Middlesex, and James E. Jeffrey. Last to board were a motley collection of Able Seamen from all parts of the world; A.T. Cooke, C. Sheppard, H. Schultz, Patrick Manyhen, J. Brown, D. O'Brien, E. Eklof, T. Shielas, W. Burke, C. Henderson, Henry Vossner, and 29-year-old Otto Johanson from Vaasa, Finland. Most of the sailors came aboard on the morning of the 17th still drunk from their final night ashore. The Star of Greece was a 'dry' ship and Captain Harrower tolerated no breach of this rule.

The second gun destined for the fort at Glenelg was to be brought out by the auxiliary steamer Guy Mannering. The ship had arrived back in London from Port Augusta with a load of wheat and was hauled into the East India Dock to discharge. The vessel was an iron-hulled, single screw, four-masted steamship, built by A Leslie and Co., of Newcastle, in 1873.

Star of Greece

Guy Mannering in Dundee Harbour, circa 1880s.
Photograph by Alexander Wilson, Dundee Public Library.

Star of Greece

She was a 2842-ton behemoth that measured 380 feet in length. The vessel was loaded at the Millwall Docks almost a month later than the Star of Greece but this was of little consequence as she would take the shorter route through the Suez Canal to arrive at roughly the same time as the crack Irish clipper.

The Star of Greece was towed from her Millwall berth to Gravesend early on March 16th to await her pilot. She had one passenger aboard, a Mr Cotter who occupied one of the spare first-class cabins. The clipper took up her tow on the morning of March 17th 1888 under greying skies and light winds the barometer indicating the rapid approach of a cold front from the west. Freezing north to northwest winds brought snow flurries and showers as the Star of Greece reached the Downs in the late afternoon to anchor off Goodwin Sands. The wind veered round to the northeast as evening fell bringing with it gales and squalls forcing Captain Harrower to make a run back to the Thames. The ship was towed into The Nore anchorage at the mouth of the Thames to wait out the nasty weather. Gales and storms blew on for another three days dumping snow all across the south of England. The crew of the Star of Greece were kept busy keeping ice and snow from the rigging as the arctic tempest blasted its way across Europe.

The winds continued to blow in from the northeast but moderated enough to allow Captain Harrower to have the Star of Greece towed again out into the Channel. The clipper took her departure from Deal on March 21st. The freshening northeaster allowed for a rapid run down the English Channel and out into the western approaches. The Lizard was passed on the 23rd as Harrower set his course to pass just west of Ushant. The winds backed around to the southwest as the clipper crossed into the Bay of Biscay. The headwinds continued until the Monte Facho light was sighted sailing past Cape Finisterre. The winds then dropped away completely and the Star of Greece lay almost becalmed for the next ten days barely ghosting along under every sail Captain Harrower could hang aloft.

A welcome change in the weather came with the arrival of the northeast trades allowing Harrower to square away for Madeira. The Star of Greece began to show her true colours as she raced south

Star of Greece

crossing the equator at 26° west, on April 16th, 24 days from the Lizard. Once into the southeast trades, fair weather prevailed to the line of the Cape of Good Hope which was reached on May 12th, 50 days out. Russell Harrower guided his ship through the great circle route searching for the prevailing westerlies that would allow a cracking run to Australia. The Star of Greece ran her easting down between 43° and 44° south with the usual westerly gales driving her across the wide expanse of the Southern Ocean.

The winds came in light and variable from the northwest as the ship passed below Cape Leeuwin. Then for the next several days light rain and moderate seas greeted the clipper as she sailed north by east towards Kangaroo Island. The overcast and showery weather continued with a quartering breeze allowing for excellent progress across the Bite. An approaching cold front on June 5th forced Captain Harrower to shorten sail as the wind veered round to the northeast and freshened bringing with it rough to very rough seas. Winds stayed fresh from the north and northeast, making for arduous sailing as the Captain ordered the ship's lookouts to keep watch for the Cape Borda light. With the passing of the trough light winds and smooth seas prevailed, the nights were star-filled and frosty as the Star of Greece glided along as another westerly change approached. Battling through rough and blustery winds and seas Cape Borda was sighted and signalled on June 10th, journeys end in sight. The Semaphore Roads were reached on June 11th at 3:30 am 80 days from The Lizard, where the ship hove to and dropped her hook to await the customs launch.

Shortly after dawn, the custom and quarantine inspectors arrived by launch from Semaphore Pier and the vessel was granted pratique and clearance to enter soon after. After laying at anchor for twenty-four hours the ship was towed around to the Port River and hauled into a berth to discharge the balance of her general cargo. With the last of her cargo removed from the 'tween deck and ballast taken on board to compensate for when the long gun was removed, the Star of Greece was shifted across to McLaren Wharf and berthed alongside the Big Crane with its 50-ton lifting capacity.

Star of Greece

McLaren Wharf, Port Adelaide, with the Big Crane in the foreground, circa 1875.
Port Adelaide Collection, State Library of South Australia.

Star of Greece

On Monday, June 25th a party from the South Australian Stevedores Company arrived and prepared to assist in the lifting of the 25 tons of gun and casing from the hold of the clipper. The main hatch covers were removed then chains holding the gun crate in its cradle were released. Soon slings and preventers were in place around the crate, and special heavy stabilising tackles were rigged to the ships masts and lower yards. With preparations in place, the Big Crane easily hoisted the gun out of the hold. The delicate operation took more than an hour as the stevedores attempted to avoid damage to the Star of Greece. Once free of her awkward burden the clipper was then warped clear and the 30-foot long crate was lowered on to the wharf to await the barge which would then be taken under tow to Glenelg.

The Star of Greece now lay tied up at the North Arm that evening to await her incoming cargo. The ship was contracted to take more than 1800 tons of wheat back to Queenstown or Falmouth for orders. There would be a delay as the twelve original Able Seamen who had signed on in London for the round trip had all deserted along with apprentice James Jeffrey. Another apprentice, Fred Skeggs had already transferred across to the Star of Erin which had left Adelaide on June 23rd. The few remaining crew of the Star of Greece could do little but clear the hatches and set up the shifting boards as they waited for the wheat laden ketches, cutters and schooners to come in from the South Australian outports. The first of the little coastal craft arrived within a day or two of the ships berthing. The loading of the grain was to be undertaken by a team of expert wheat lumpers. These men whose sole job was the shipping of wheat into the holds of arriving clippers worked twelve-hour days at times but stopped dead on 6:00 pm. Returning to shore for the next load in the morning.

The steamer Guy Mannering meanwhile had left London on April 18th outbound for Adelaide, stopping at Liverpool to take on the outward mail and passengers. Sailing via Marseilles, Malta, Port Said, Suez, Aden and onto Perth then Adelaide she steamer arrived off of Semaphore on June 25th, 68 days from London. The steamer was brought in alongside McLaren Wharf on the morning of July 4th and the 25 tons of gun and casing and 100 tons of military equipment were hoisted out of her hold onto the wharf. Labourers from the South Australian Stevedore Company then removed much of the equipment to a shed that been erected on the wharf to house the guns until they could be transhipped by barge to Glenelg.

When Captain Harrower was not aboard the Star of Greece attending to the needs of his ship and her remaining crew he was ashore dealing with the ship's agents and the Seaman's Boarding House runner attempting to recruit at least twelve fresh hands for the run home. He was also missing the services of two much-needed apprentices. Advertisements and notices had been published in sailor's homes and Mercantile Marine Offices in both Victoria and New South Wales offering foremast hand positions at the rate of £6 per month plus a sign-on bonus. Most sailors in Port Adelaide were demanding £7 a month and this was well beyond the rate of payment approved by the Star of Greece's owners back in London. Russell Harrower was having to accept a motley assortment of foremast jacks of all skill levels and experience. Henry Crowe the boarding house runner had to work hard to earn his per head commission. Yet as the ships departure date grew closer there were few hands available to fill the fo'c'sle and even Russell Harrower's brother-in-law, James Bishop, a clerk working for the Port Adelaide Steamship Company was keeping an eye out for likely bodies to fill the ships manpower needs.

Life was not all work for Russell Harrower though, his sister Alice and her family resided just a short train ride away from the port at Woodville. In the ten years since the two siblings had last embraced the Bishop clan had grown considerably. James Bishop was doing rather well for himself as a shipping clerk and assistant broker. He was a keen yachtsman and owned a part share in the 15-foot cutter Xanthe which regularly raced at Glenelg. Alice Bishop and her brood were frequent visitors to the Star of Greece and many an hour was spent by the children aboard the docked vessel. Often accompanying Alice was a rather unruly gaggle, John aged 14, Fred, 13, Frank, 9, James, 7, Allan, 4, and little Mabel aged just 2. When not entertaining friends and family aboard ship, Russell spent his nights in Woodville at the Bishop household catching up on a decade of lost moments. Firsthand news from home was most welcome and many family stories were shared around the kitchen table as the children either sat rapt with wonder at their uncle's tales or roamed noisily around the house and garden.

Portrait of Alice Bishop (nee Harrower).
Randall Family Archive. State Library of South Australia.

Star of Greece

Despite not having children of his own, Uncle Russell's gentle ways made him a firm favourite with his nephews, and perhaps like the tales told of his grandfather, he too related somewhat of his past adventures at sea. There was genuine and familiar warmth to be found within the Bishop home. It was a snug harbour where Harrower could relax and for a few hours at least, forget about the lonely responsibilities that came with being a deep-water ship's master.

Loading of the Star of Greece was continuing apace as more wheat came in from the outports, the lumpers keeping up a steady flow of wheat bags into the clippers three holds. As the first few thousand bags had gone down the slides John Hazeland and the carpenter Robert Donald set the apprentices into the piles with knives. There deep in the bowels of the lower hold they set to slashing open many of the outer bags to allow the wheat to flow into the hollows and corners against the shifting boards. The wheat would eventually settle down to the cracks and spaces allowing the load to settle evenly. Soon enough the loose grains would weep into the bilge and begin to ferment creating a toxic black sludge that would fill the lower hold with a nauseating miasma. Removing this toxic sludge would be a job for the apprentices to attend to once they arrived back in London. The clipper had several dozen tons of red brick rubble aboard as ballast for the wheat being relatively light made the clipper rather crank. The heavy ballast was needed to allow the Star of Greece to be successfully trimmed for the run home.

Foremast hands were slowly gathered from across the southern colonies and from ships that had recently arrived in port. Two mates, Alf Organ, 35, and John Airzee, 28, previously aboard the ship Avenger, had come across from Melbourne a week after the Star of Greece had docked at Port Adelaide. The two were experienced sailors and were quickly snapped up by Mr Crowe as they sought lodgings at the Sailors Home. Neither man was in any way unusual for a Foremast Jack of the time, thus it was no surprise to anyone when John Airzee was arrested for being drunk and using offensive language in Vincent Street, Port Adelaide on July 1st.

Star of Greece

The barque Beltana moored at Port Adelaide, circa 1880s.
A.D. Edwardes Collection, State Library of South Australia.

Star of Greece

Spending the night locked up in the local watch-house the badly hungover sailor appeared before the local magistrate the next day and was fined 20 shillings for his troubles. His drinking partner, Edward Percival had given the local constabulary a lot more trouble for he was fined a total of £2 15 shillings, a considerable amount for a sailor without a berth.

One man who signed articles early in the piece was Peter De Smet, from Ostend in Belgium. The son of a widowed storekeeper, 39-year-old Peter was an experienced old shellback who had come into Port Adelaide aboard the 900-ton barque Elizabeth Nicholson, a timber drogher from Vancouver, via New York. Her wages were half that offered by Captain Harrower and having quickly spent his meagre earnings in the dockside pubs and brothels. De Smet was delighted to sign on for £6 a month.

Another of the prospective crew was a 21-year-old Swede named Carl Claeson. The young adventurer had arrived in Port Adelaide back on May 8th aboard the 741-ton barque Beltana from London and had decided that he had experienced enough of the danger of a life at sea. So jumping ship he hot-footed it out of Adelaide to avoid being imprisoned for deserting his vessel. Claeson found himself in in the countryside near Murray Bridge working for a young family of German migrants. The couple with three young children were working hard to clear the scrub from their block and had employed the wayward sailor as a labourer. Everything they grew was sold to pay for the development of the farm and Claeson was forced to sleep in a ramshackle barn in the depths of a cold, wet winter. Carl stuck it out for a few weeks but then decided that the work was not for him. He was soon wistfully yearning for the relative freedom of a life at sea. Carl caught the train back to Adelaide and before long washed up at the Sailors Home. Henry Crowe soon found him a berth aboard the man-hungry Star of Greece. The Prince Alfred Sailor's Home was the chief point of recruitment for Captain Harrower who was in desperate need for still more experienced foremast hands as the day of departure loomed.

Star of Greece

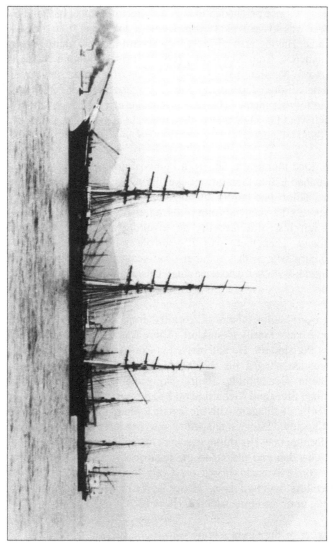

The iron ship 'Greta', 1249 tons, circa 1885. Built in 1874, William Lowden and Co.
A.D. Edwardes Collection, State Library of South Australia.

Star of Greece

Amongst those persuaded to sign articles for a run aboard the Star of Greece was 43-year-old George Irvine, a dour Scott from the Shetland Islands. Having arrived in Sydney aboard the ship Donnelly, he had come across from Newcastle, N.S.W., after sighting a notice in the Seaman's Mission there. George, the son of a crofter had a wife and several children dependent upon his wages and he was anxious to get home to visit them. The Star of Greece offered a substantial wage which would be most welcome in the hand to mouth existence of most Shetland Islanders.

The three month trip aboard a homeward bound clipper would put more than £20 in George's pocket upon the ship's arrival in London. Other sailors had family members reliant upon their pay for survival. There was 50-year-old John Gatis, a father of four adult children who had left his wife Isabella, an invalid, back home in Montrose, Scotland. Then there was 27-year-old Wilhelm Oerschmidt, from Hamburg who with his father dead sent much of his pay home to support his mother who ran a small Gästehaus in the city of Wilhelm's birth.

Not every sailor who was recruited from the Sailors Home by Mr Crowe had a family to support. There was 19-year-old Henry Corke from Portsmouth. He had arrived at Port Adelaide in mid-May as an apprentice aboard the 1200-ton ship Greta from Hamburg with Wilhelm Oerschmidt. Henry along with two other apprentices William Steel and Richard Scott had deserted the vessel on June 11th, rather than sail again with the legendary hard driver, Captain William D. Cassady. Henry managed to stay free despite the best efforts of the local police. His shipmates were not so fortunate and were quickly apprehended and placed in the local lock-up. The two young men were ordered back aboard the Greta their brief run ashore having lasted less than two days. Henry Corke eluded capture remaining in hiding until the Greta sailed on June 15th.

Another of the Greta's crew who found his way aboard the Star of Greece was 24-year-old Eduard Bluhm from Libau, Latvia. His family was originally from Potsdam but had moved to the port city where they dwelt in a modest home on Kronenstrasse. Eduard had deserted the Greta along with much of her crew and was eager for a

Star of Greece

higher paying berth away from the tender mercies of Captain Cassady. The last of the crew to be signed on was a collection of experienced Able Seamen; William James Miles, 31 from Middlesex who had signed off from the Star of Erin, David Bruce, 24 from Scotland, who had come to Port Pirie aboard the barque Edinburgh, and James T Revell who had arrived from Fremantle aboard the barque Raven.

By Monday, July 9th 1888 the last of the 16002 bags of wheat was ready to be loaded aboard the Star of Greece. The regular team of wheat lumpers employed by John Darling & Son arrived at North Arm early, ready to begin their final day's work. Albert Pearce who was the chief agent for the company and supervisor of the stevedore gang boarded the ship to be greeted by first officer Hazeland. The two men watched as the canvas covers were removed from the main hatch. Across the reach the crew of South Australia's only warship, HMS Protector were busy drilling and maintaining the upkeep of the doughty little dreadnought. One of her crew, F.J. Dawson watched the last bag of wheat being taken aboard as the crew of lumpers gave a mighty cheer celebrating the finish of their work. The ship's Steward George Blackman came back from town that afternoon accompanied by his assistant George Carder. Together the two men supervised the loading of fresh food and stores by several members of the crew. The supply wagon rang with the cackles of a dozen chickens and the squeals of several pigs which came aboard ship in crates. The chickens were placed in the twin coops situated abaft the fire buckets atop the break of the poop and the pigs were lead squealing to the sties placed in the shelter of the fo'c'sle head.

Captain Harrower came back aboard that evening accompanied by his sister Alice, her husband James and their large brood of children. There the family spent a long and happy evening seated around the Star of Greece's substantial saloon dining table sharing a last meal. With no cook aboard it was left to George Blackman and George Carder to provide the evening's repast. They were joined by the ship's officers John Hazeland, Bill Waugh and Charlie Commerford, all of whom were poignantly reminded of their own families and the warmth and comforts awaiting them at home.

Star of Greece

Prince Alfred Sailor's Home, Port Adelaide, circa 1878.
Sweet, Samuel White, 1825-1886, State Library of South Australia.

Star of Greece

With the evening finally over the Bishops left the ship with their very tired offspring in tow. It was planned that Russell would join them at the family home in Woodville for luncheon the next day.

Tuesday morning saw Captain Harrower ashore early having left instructions with Hazeland to prepare for the clipper's move out to the Semaphore anchorage. Russell still had business ashore to take care of and the mate was more than capable of taking the ship to anchor in the roads. His first stop was to the Harbour Master's Office, where he arranged for a local pilot, Captain Alexander Boord to oversee the shifting of the Star of Greece.

He then arranged for a tug or two to shift the heavily laden clipper down river and out to sea. Then after making a few stops to gather last-minute personal gifts and supplies Russell made his way over to the port offices of John Darling & Sons to get customs clearance and to check on final crew numbers registered with the office. As he was filling in the cargo and stores declarations one of the shipping clerks commented;

"Well, Captain you're not going to get drunk on the way home, as I see you have only a little spirits onboard for medical purposes. No, replied Captain Harrower, *I make it a rule never to touch spirits aboard ship, and I have had that little lot on board since we left England."*

Appearing on one of the bills of lading were several cases of wine he declared were a gift for his mother Jemima back in London. These were amongst the last of the cargo to be stowed and were carefully hidden by George Blackman to keep them away from an ever-thirsty crew. As Russell was seeing to the ship's business ashore, the Steward and George Carder were out and about the local victuallers gathering up the last of the cabin stores, a few choice delicacies, sparkling waters and fresh food for the run home. George Blackman was on the hunt for a few treats for his own family, having left a wife Ellen and two young children back home in London. The two men along with several of the crew including James Revell were out for one last run ashore before spending the next three months at sea with few if any comforts and no beer or grog to speak of.

Star of Greece

Having finished his business for the day Russell caught the train from Port Adelaide to Woodville to spend the afternoon and evening with his family. The sky was coming over cloudy as the cotton ball-shaped altocumulus wandered in from the west, early indicators of a blow and rain in the offing. The quiet and unassuming Captain was not one for great shows of public affection but this did not extend to his sister and her boisterous crew. The warmth of his welcome at the Bishop home was equalled only by the love they all shared. The evening was spent with a last meal and final farewells for the children, Uncle Russell would be off in the early morn long before the children awoke from their beds.

Back at the North Arm Wharf John Hazeland and the pilot, Captain Alexander Boord were making final preparations to have the Star of Greece hauled out. The last of the crew were expected to soon board however Mr Crowe came with some bad news. One of the expected hands, an Able Seaman named Harry Edwards, was not going to turn up any time soon. He had been arrested for stealing a watch and was currently appearing before the local magistrate. There was some good news on the horizon for James Bishop was seen approaching the ship with a forlorn trio of men in tow. Once aboard James Bishop explained that whilst they were unable to pay for their passage back to Blighty the men were more than happy to work for their passage home. Hazeland informed Bishop that it was not for him to employ the three newcomers, but he would put their case to Captain Harrower once he came aboard. The youngest of the three prospective deckhands was 19-year-old Robert Muir from Glasgow. The eager young adventurer was out to see the world. His father had paid £22 for his son to travel to Australia aboard the Loch Katrine. Robert had arrived in Melbourne back in April and now broke was looking for a way home and the Star of Greece presented such an opportunity.

The other two offered much less information about themselves, the elder of the two, 40-year-old John McVicars stated that he and his companion Andrew Blair had been working as props men at the Royal Theatre in Adelaide. The two were shabbily dressed but McVicars came with a large steamer trunk marked with the name Stevens.

Pilot, Captain Alexander Boord.
State Library of South Australia.

Star of Greece

What James Bishop did not know or failed to mention was the fact that both men were wanted for questioning by police over a series of thefts that had taken place at the Royal Theatre in the preceding weeks. John McVicars who also went by the name John Rowlyn was well known to police having dodged a fraud conviction for presenting dud checks earlier in the year. The man may have also been somewhat mentally unstable having been convicted of an unsuccessful suicide attempt in Queensland in 1886. Living a hand to mouth existence McVicars and Blair were currently on the run and looking for a quick way out of Adelaide.

A steam tug came up alongside the clipper, lines were taken in and roved to the bits, and then as the tug's whistle sounded its mournful wail moorings were cast loose by hands ashore. Once out into the stream another tug pulled ahead as her tow line came on taught. John Hazeland ordered the inner jib set and fore and main lower topsails brailed up to assist the tugs in shifting the clipper downriver. The weather was cold and blustery as under the direction of the Captain Boord the Star of Greece was guided out through the outer harbour and down to the Semaphore roadstead to await the return of her captain. Once at anchor Captain Boord had little left to do but wait whilst John Hazeland ordered the crew aloft to bend the remaining sails. Aloft, sailors made sure to check the shrouds, stay, gaskets and ratlines for any loose, strained or rotten lengths. Any found were immediately cut away and overhauled. During the afternoon a steam launch came out from Semaphore Pier with another three sailors, bringing the total up to eleven.   By days' end the foremast hands who had boarded the Star of Greece were Able Seamen; John Airzee – 28, John Gatis – 50, Peter De Smet – 39, Carl Claeson – 21, Alfred Organ – 40, George Irvine – 43, Wilhelm Oerschmidt – 27, David Bruce, William Miles – 32, Eduard Bluhm – 24, and signed on as an Ordinary Seaman, Henry Corke – 18. The fo'c'sle now had almost its full complement, a mixture of young boys, seasoned sailors and hardened old shellbacks.

Russell Harrower spent Tuesday night and most of Wednesday ashore taking care of the last of the ship's business and spending time with his family. With the children being cared for at home Alice and James accompanied Russell aboard the train to Semaphore.

Richard 'Dick' Jagoe.
State Library of South Australia.

Star of Greece

Dick Jagoe's fleet of passenger steamers, and pilot cutters.
State Library of South Australia.

Star of Greece

The weather at the time was cold with squally showers coming in from the northwest. Winds were variable wafting around to all points of the compass, the glass slowly falling. As Captain Harrower completed the final paperwork at the Semaphore Customs Office he was strongly advised by Health Officer Richard 'Dick' Jagoe not to head out to the Star of Greece as it was too rough for the launch to safely motor out to the anchorage. The churning grey-green waters off the Semaphore Pier were indeed a daunting sight thus heeding Jagoe's advice the trio headed home to Woodville to wait out the approaching storm. Before leaving Russell arranged with the Dick Jagoe for his launch to be ready to take him out to the Star of Greece first thing in the morning.

Wednesday also saw the arrival of the ships papers and the final crew member, James Revell, 40. The old dishevelled sailor had signed papers on Tuesday afternoon and took the launch out to the Star of Greece at 10:00 am on Wednesday along with Henry Crowe, George Blackman and young George Carder. Revell was quite drunk when he arrived and had to be helped aboard, having spent the previous evening carousing in the local pubs. With a full complement aboard Hazeland could finalise preparations and the sailors could be assigned to their respective watches.

Thursday, July 12th 1888 dawned the skies appeared a tad overcast and the winds were light and at times blustery from the northwest. Captain Harrower and the Bishops arrived at the Semaphore Pilot's Station soon after breakfast, there to meet the launch prearranged by Russell the evening before. Dick Jagoe was out and about to oversee the loading and departure of the small craft and to handover, the newspapers the Star of Greece would carry back to London. With a cheery toot, the pilot's launch motored out to the roadstead, an anxious Captain and his family aboard. Once again on his ship Captain Harrower was greeted by the officers and the steward who was there to offer refreshments to the new arrivals. Both port and starboard watches were busy readying the clipper to get underway. With Dick Jagoe ready to depart James Bishop said his farewells to Russell and Alice, promising to return late in the afternoon to see the ship and her captain off. The little steamer chuffed her way back through the chop to Semaphore to deposit Mr Crowe and Captain

Star of Greece

Boord back to shore. Alighting from the launch James said his goodbyes and with a quick backward glance headed to Port Adelaide and his job at the offices of the Port Adelaide Steamship Company. Alice Bishop stayed aboard the Star of Greece to spend the last day with her brother.

There had been a rather animated discussion aboard the Star of Greece before the pilot departed. John Hazeland was anxious to get underway as soon as possible, the weather was forecast to worsen as the day progressed. Alex Boord, on the other hand, cautioned Captain Harrower to await the flood tide that would come late in the afternoon. A final if, unconvincing decision was made by Russell Harrower to wait for the rise of the tide whilst keeping a weather eye out for any sudden drop in barometric pressure or change in the wind. With the pilot gone Captain Harrower set to putting his vessel in order. It was once the launch had left that the mate presented the three men wanting to ship home as working passengers. Russell recognised one of the trio as an individual who had approached him, using a different name, whilst Star of Greece was docked at the North Arm looking for third-class passage home. When Captain Harrower had told him that they had no such berths, the rather unsavoury looking gentleman, John McVicars had gone away deeply disappointed. Now here he was, technically a stowaway, but with the Star of Greece being shorthanded Russell begrudgingly signed the men on. Trusting his brother-in-law's judgement the trio of vagabonds were put on the books as deckhands for 1 shilling a month plus food. -

Upon being questioned further it turned out that Andrew Blair was a sailor of some experience and so was put on the books as an Able Seaman. He was ordered to place his dunnage down in the fo'c'sle and Henry Corke was moved back to the half-deck with the rest of the apprentices. Robert Muir was sent aft to help George Blackman in the galley and McVicars was given a berth in the mid-deck with the ships idlers. Thus the Star of Greece had enough men for the voyage home; Captain Harrower 29, First Mate John Hazeland 24, Second Mate Bill Waugh 23, Third Mate Charlie Commerford 21, Boatswain Bill Parker 41, Carpenter Bob Donald 27, Sailmaker Gus Carlson 61, Steward George Blackman 36, Cabin Boy/Cook George Carder 19 and four apprentices, Frank Kearney 19, Ed McBarnett 18,

Star of Greece

Alf Prior 19, and James Johnstone 19, plus the new hands. In all twenty-eight men and boys prepared to make sail to Falmouth for orders.

With 1840 tons of wheat and 200 tons of ballast aboard the Star of Greece was riding well down to her marks in fine trim. As the day wore on the crew were aloft bending the last sails and overhauling the rigging in readiness for the voyage ahead. Captain Harrower left the last-minute jobs to his officers whilst he checked over the ships paperwork, crew manifest, bills of lading and store lists. A constant eye was kept upon the barometer and the sky above. The morning had started cloudy but fine, then towards lunchtime, the skies began to darken, the wind picking up. The smell of moisture-laden dust was in the air as the wind strengthened through the afternoon from the northwest. Russell was still in two minds about setting sail on Thursday evening, with the build-up of threatening clouds from the west. Alice urged her brother to stay his hand and wait out the coming tempest. Having lived in Adelaide for the last decade she was well aware just how quickly weather and sea conditions could change.

Hazeland kept on at his captain, insisting that they leave on the evening tide. He argued that all the sails had been bent and the crew had completed preparations. The ship and crew were more than ready to depart. The pilot launch was seen to be approaching the ship just before 4:00 pm signifying that Russell would soon have to make a final decision. To go or not to go? That was the question.

The launch was soon moored at the base of the gangway and stepping aboard was James Bishop, Dick Jagoe, and Customs Inspector Arthur Searcy. Harrower greeted his guests with all due respect as these port officials would give him his final health and customs clearances needed to depart. The various manifests, logs and bills of lading were examined and stamped according to port regulations. Arthur Searcy took careful note of the deteriorating weather conditions and urged in the strongest terms for Captain Harrower to delay his departure for a further 24 hours, echoing the concerns of his sister Alice. Yet the arguments of his first mate and the continuing relatively fine conditions convinced Russell that the exhortations for delay were a gross overreaction.

Star of Greece

The Signalling Station at Semaphore on the coast of South Australia, circa 1877The two-storey house in the background was owned by Richard Jagoe, newspaper shipping reporter, quarantine officer and launch fleet operator.
State Library of South Australia.

Star of Greece

Semaphore Pier, circa 1890.
State Library of South Australia.

Star of Greece

With confidence he did not feel the captain explained to his guests that in times past he had survived several Indian Ocean cyclones and a partial dismasting on his first trip aboard the Star of Greece. Thus the predicted blow held little to fear. The wind was blowing fresh to strong from the north-northwest and the barometer was holding steady at 30.00; almost perfect sailing conditions for a quick sprint down Spencer Gulf and out through Backstairs Passage. The time came for final farewells at 5:30 pm when Russell and Alice bid each other a tear-filled goodbye. Neither knew when each would see the other again but there was hope that their parting would be brief. Russell Harrower farewelled his brother-in-law, James Bishop, Arthur Searcy and Dick Jagoe as they boarded the launch that would take the party back to Semaphore.

As the little steamer bounced her way to the pier Captain Harrower gave orders to prepare the Star of Greece for getting under weigh. John Hazeland moved to the fo'c'sle head taking with him members of his watch with orders to bring up the anchor cable until it was hove short. With the clipper's head to the wind, it was necessary to send crew out to onto the bowsprit to be ready to set the inner jib. From the poop, Russell alternated between waving farewell to his sister and issuing orders to his officers and crew. One of the more experienced apprentices was sent to the wheel alongside David Bruce in preparation for when the anchor was tripped. As the sun set, a lurid reddish glow filled the horizon below the gathering clouds. At the same time, the port and starboard side-lamps were lit by an agile apprentice. The wind was light to moderate from the west-northwest with little chop to the sea. The glass was beginning to fall, now down below 30 with no chance of it rising anytime soon. Captain Harrower gave the order for the anchor to be hauled up at 6:15 pm. The men pushing hard on the capstan bars raised ten fathoms of cable before the anchor was hove short. As the anchor broke loose from the bottom Russel Harrower gave the order to set the foretop staysail. The halyard was then sheeted home allowing the sail to catch the wind. The bosun Bill Parker drove the sailors on the capstan hard with orders to fish then cat the hook as the clipper's head came round to the wind.

Star of Greece

Sailing on to Infamy

A tug leaves her charge off Semaphore, circa 1900.
State Library of South Australia.

With a quartering northwester, Bill Waugh bellowed for his watch aloft to set reefed lower fore and main topsails. The sailors scurried out upon the footropes as the clipper swung into position. At 6:45 pm Harrower gave the order to loose gaskets and the heavy upper topsails ballooned out as they caught the strengthening wind. The Star of Greece began to make headway slipping quietly through the increasingly choppy waters as the yards were braced around to make the most of the wind as lines were sheeted home. The mate's watch worked feverishly to stow the anchor even as the vessel was getting underway. The taught wires of the standing rigging began their haunting song as the ship gathered speed. The vessel canted hard over on the starboard tack as Captain Harrower set the course south by southwest magnetic, towards Cape Jervis. Just as they were gathering speed the mizzen topsail halyards carried away when the crew were hauling up the mizzen topsail yard. The pennant carried away and the topsail yard came down crashing atop the cro'jack whilst at the same time several blocks came hurtling to the deck.

193

Star of Greece

The Star of Greece running before the wind.
Brodie Collection, State Library of Victoria.

Star of Greece

James Revell was not quick enough to get clear and was struck on the head knocking him senseless, leaving a nasty gash on his forehead. The luckless sailor was carried below and laid in a spare cabin to be tended to by the Steward. With the ship making steady progress Captain Harrower and John Hazeland took the time to compare and calibrate the twin chronometers. They then checked the log, the ship making about 6 knots, the yards almost square. Orders were then given for the duty watch to set the main upper topsail which was speedily done. It being the middle of the dog watches the crew were sent to tea, firstly Bill Waugh's men and then the sailors in the port watch. The wind began to pick up about two miles into the run south and Russell invited Hazeland below to check the ship's charts. Emerging back on deck Captain Harrower ordered the upper topsails and mainsail clewed up and reefed. Russell handed over command to Hazeland and went below to have dinner before turning in for an hour or two.

Having been served a light dinner by George Carder, Russell retired to the settee in the saloon to try and get some sleep. Checking the barometer before laying down he did not bother to undress or remove his sea boots knowing at a glance of the falling glass and pitching and rolling of the ship that those aboard were in for a rough night. At 8:30 pm a northerly squall overtook the ship and Hazeland ordered Charlie Commerford forward to rouse the duty watch with orders to furl the upper topsails and make them fast in the gaskets. The grumbling sailors emerged from the shelter of the fo'c'sle in their wet weather gear and scrambled aloft to make the sails secure as the Star of Greece began to heel over in the increasing gale. Awakened by the sudden movement of the ship Captain Harrower roused himself from the settee and after gaining his bearings headed up the stairs leading from the saloon to the poop checking the barometer, and chronometer on the way. Emerging into the midst of a ferocious squall he hauled himself across to the windward binnacle and checked the course. The Star of Greece being heavily laden was shipping great amounts of water over her lee rail making the deck a difficult and dangerous place to work upon.

Harrower noted an increase in the clipper's speed despite being under shortened sail. He did not want to run the risk of attempting

Star of Greece

Backstairs Passage at night. Whilst there were lights at Cape Jervis and Cape Willoughby the strait was rather narrow with treacherous currents, cross seas, hidden reefs and shoals. Thus he came to a difficult decision, realising too late that he should have listened to the sage advice of Arthur Searcy and prescient words of his sister.

At 9:00 o'clock Russell ordered the port watch to clew up the topsails and brace the main yard aback. He then ordered the watch to set the mizzen staysail and for the helm to be put up to bring the ship back on course. The Star of Greece was cracking along at more than 6 knots heeled hard over on the starboard tack when she was hove to. The clipper still managed to make 1-2 knots steerageway as lines of freezing squalls roared in from the west. Once the sails had been made fast and things squared away Captain Harrower ordered the deep-water lead to be used and thence every quarter-hour after. He checked the compass once more before going below. A worried Captain left explicit orders for John Hazeland to call him immediately anything unusual occurred.

Unable to sleep nor shake the feeling that something was amiss Russell was soon back on deck, the wind having veered right around to the southwest, blowing even harder. With the change came enormous high peaked rollers that battered the clipper's side and pushed the Star of Greece hard over to port. The luckless lookout had to be lashed to the fo'c'sle head as the wind and waves threatened to hurl him overboard. The order was given by the Mate to keep a sharp lookout for lights and land on the lee bow. Unable to take a star sighting Hazeland was oblivious to the fact that the Star of Greece was drifting at almost 5 knots to leeward, ever closer to danger.

By 11:00 pm the wind was blowing gale force from the southwest bringing with it rain, hail squalls and a high cross sea. Captain Harrower was on deck every thirty minutes or so and kept the lead going. At 11:30 the twinkling beacon of the Port Adelaide Lightship was sighted to the north-northeast, allowing Captain Harrower to gain an accurate reading of the clippers position and course. He informed Hazeland that the Star of Greece was on track and travelling well though the lookouts were to maintain their vigilance and the lead was to be swung every quarter-hour. Just before midnight, the deep-sea

Star of Greece

lead was run out with 60 fathoms of line, yet no bottom was met. What the sailors could not know was the amount of leeway the clipper was making towards the shore.

At eight bells the port watch was ordered below for a well-earned rest. Bill Waugh and the starboard watch came up on deck into the teeth of a howling Southerly Buster, dressed in their slickers, to receive orders from the mate. He and Russell Harrower were both on the poop, however, the captain was happy to allow the first officer to relay the commands to Waugh whose watch it now was. The incoming helmsman was ordered to maintain the ships' course south by southwest and then Bill Waugh ordered the lead maintained. Just before retiring Russell gave instructions to brace up the foreyards. Going below he heard the dulcet tones of the boatswain William Parker bellowing out to the starboard watch to bring the foreyards around to port to try and again bring the clipper round onto the correct heading. She was slewing to leeward in the trying conditions, her rudder failing to bite.

Once in the cabin, Russell took note that the glass had now dropped to 29.64 and was continuing to fall. Hazeland remained on deck for another thirty minutes to make sure that all was well. A sudden gust of wind filled the backed main topsail and threatened to tear it right out of the boltropes. Waugh ordered James Johnstone and other members of the watch aloft to spill the wind out of the sail and haul the topsail yard around. At 1:00 am Harrower was back on deck and fearful that the Star of Greece was drifting too close to shore commanded all the yards be swung round onto the starboard tack. He then ordered the spanker set five feet by the head, and for the helm to be put half down to keep the clipper on what he thought was a steady course south by southeast towards Backstairs Passage.

The dirty weather continued through the night giving those below little chance of rest. Waugh ordered apprentice James Johnstone aloft to the foretop to keep a sharp lookout for land or lights on the lee shore. Soon after the call came shouted down that James thought he had sighted land off the port bow but due to the thick weather and coal-black night Bill Waugh did not think that this could have been possible. Thus he did nothing further with the sighting. This was at

Star of Greece

1:30 am on July 13th 1888. Carl Claeson took over the weather helm at 2:00 am with orders to keep up to the wind as close as possible. Apprentice Alf Prior took over from James Johnstone in the foretop with the same orders to look out for shore lights and land to leeward. The earnest young man was faced with an almost impossible task with visibility down to under a mile and frequently reduced to less than one hundred yards thanks to hail squalls sweeping in from the west. As the ship thrashed her way south Captain Harrower was on deck every ten to fifteen minutes glancing at the binnacle and the patent log. Soon after he ordered the spanker taken in and kept all sails furled tight on the mizzen-yards. He then returned below to rest having not slept properly since the night before.

At 2:30 am the helmsman Carl Claeson shouted out to Bill Waugh that he had sighted land close on the port bow. This sighting was confirmed by Alf Prior in the foretop and then almost immediately the second mate sent an apprentice below to call up the captain and mate. Jumping up from the settee, still in his wet weather gear, Russell raced up the companionway to be met by Hazeland. The two men reacted with cool immediacy upon seeing the horrendous danger the Star of Greece was in, caught in irons upon a lee shore. Captain Harrower gave the order for 'All hands on deck!', and for the first mate to immediately prepare the port anchor and 60 fathoms of cable for release. John Hazeland raced forward letting out three blasts of his whistle calling up the port watch from below. The sailors scrambled out from their bunks struggling to dress as they assembled by the fore hatch.

Wasting little time the crew sprang up the ladders to the fo'c'sle head and hauled the anchor into position. At 3:00 am the starboard watch under Bill Waugh was ordered to brace up the yards and heave the lead as the clipper drifted closer towards the shore. In short order the port anchor was over the side and let go, the cable running out to windward. Unbeknownst to those aboard the Star of Greece the cable was wrapping itself around the anchor flukes and stock preventing the hook from gaining any traction in the sand and mud below. A heavy sea was running and water continued to be shipped over the weather rail as the clipper drifted shoreward.

Star of Greece

The Wreck of the Star of Greece. 13th July 1888
Postcard by A. Dufty, circa 1888.

Star of Greece

Captain Harrower and those with their eyes turned to the shore could see the white foam and hear the dull thud of breakers fast approaching to port. The ship drifted helplessly for little more than five minutes before she found herself in the surf zone less than four hundred yards from the beach. Spume topped rollers repeatedly lifted the helpless clipper high upon their crests bouncing her up and down in the increasingly shallow waters of Aldinga Bay. Miraculously the ship slipped by the outer reefs of Port Willunga and Lions Head, rocky teeth that would have torn the bottom out of the vessel if she had struck there.

Instead, the hand of Neptune carried the Star of Greece into a narrow gap between Lion Reef and the Headland overlooking Gull Rock. The wind, waves and current drove the clipper bow first into the sandbank two hundred yards from the shore. The ship came to a shuddering halt as she buried her forefoot deep into the mud and sand before the grounded vessel was pushed broadside to the beach. The wheel spun freely as it was torn from Carl Claeson's grip. Soon the Star of Greece began to cant over to starboard her weather rail almost underwater. The decks of the doomed ship were exposed to the full force of the waves as they rolled in from the storm-ravaged gulf.

Russell ordered Hazeland to send his men aloft to clew up the fore and main topsails. This helped take the pressure off the ship and prevented her from rolling right over. The port watch acted with calm efficiency to comply with the orders as great waves of green-white water flooded across the deck. The starboard bulwarks began to be carried away making the work all but impossible. Harrower then sent several men below to break out the signal rockets but upon coming up onto the poop the rain tore at the wax paper the rockets were wrapped in, soaking them through. John Hazeland then sent the men below again to fetch out the twenty blue distress flares from his cabin stores. The great dollops of seawater sluiced across the deck making work forward all but impossible for Bill Waugh and his crew. They were forced to seek shelter within the fo'c'sle.

Those sheltering within the cramped confines of the forward crew quarters were William Waugh, Gustav Carlson, Frank Kearney, Ed McBarnett, James Johnstone, John Airzee, John Gatis, Peter De Smet,

Star of Greece

Alfred Organ, George Irvine, Wilhelm Oerschmidt, David Bruce, William Miles, Henry Corke, and Eduard Bluhm. Just after 3:00 am Captain Harrower ordered Charlie Commerford and Boatswain William Parker aloft to the mizzen top with the blue flares, joining them were Robert Donald, Alf Prior, James Revell, Andrew Blair, Carl Claeson, Robert Muir and John McVicars. The blue lights were lit in pairs, yet as the night wore on, and the ship threatened to roll all aboard into the surf, there was no response from the dark, storm-battered shore. Russell Harrower, James Revell, George Carder, George Blackman and John Hazeland stayed put in the companionway, knowing that there was nothing more that could be done until daylight.

The men in the mizzen rigging continued to burn blue distress flares at intervals throughout the night to no avail. No lights could be seen from the shore and the howling wind, the thump of the surf, and pouring rain made communication with those on deck all but impossible. The men sheltering fore and aft spent a terrifying night trapped in the fo'c'sle, cabin or the aft rigging waiting for the sun to rise. By now the starboard bulwarks had been stove in and the decks swept clean of everything moveable. The first casualties of the calamity were the three pigs washed away with their sty. The luckless animals disappeared over the side along with the harness casks, water barrels, spare spars and anything moveable. Before sunrise, the mid-deck and half-deck doors were stove in and their contents washed out. The galley, bosun's locker, carpenters' workshop and sailmaker's store were gutted by the raging sea.

With each wave slamming into the Star of Greece the stern section vibrated madly and deck planking began to crack and splinter. Then at 4:30 am a huge comber rolled over the poop and with the scream of sheering metal, carried away the saloon skylight and stove in the booby hatch that granted access to the steward's cabin abaft the mizzenmast. The cabin was flooded up to the wastes of those sheltering below knocking the Captain, Revell, Carder, Blackman and Hazeland off of their feet. Russell Harrower ordered them up the stairs and back into the companionway.

Star of Greece

Trapped in the companionway the five men were left with few options. Harrower and Hazeland ventured out onto the poopdeck a little after 5:00 am to ascertain their position and the condition of the ship and weather. In the predawn darkness, neither man saw the oncoming combers, these waves larger than the last. Before they could react boiling waters slammed into the side of the clipper causing it to rock wildly on its keel. The first waves swept over the poop carrying away the hencoops, poop railing, fire buckets and both men. Captain Harrower was swept aft into the taffrail whilst the first mate disappeared off the poop and was pushed forward with the flooding waters. Picking himself up a dazed John Hazeland staggered forward to find shelter with the other men within the fo'c'sle. Once through the hatchway, he found it filled with petrified sailors the floor awash.

Aft the Captain found himself being hauled from the rail by George Blackman who helped him once more into the shelter of the companionway. The stern section shuddered chillingly with the shock of each wave's impact and the upper masts swayed dramatically with each roll of the ship. Another series of large breakers crashed into the clipper carrying away the forestays. Like giant toothpicks, the wooden royal masts snapped and came crashing down, bringing with them the royal and topgallant yards. The main topgallant yard speared down through the aft cargo hatch with a mighty crash splintering wooden battens and covering boards. The cro'jack collapsed in its slings as tons of water began to pour into the aft cargo hold. With each new wave more seawater flooded in below. The normally dry bags of wheat quickly soaked up the water and soon began to swell filling up every open space within the already tightly packed hold.

Those in the cabin could feel the hull creaking and twisting as the cargo began to shift and swell. The swollen grain put great strain on the already weakened strakes and bulkhead below. The thundering surf battered the grounded ship and Captain Harrower knew that it was only a matter of time before the remaining masts and yards came plummeting down upon the deck. At about 6:00 am the spanker boom broke free from its lashings and the ropes keeping it aloft gave way. The freewheeling spar slashed wildly from side to side before it finally came plummeting down upon the companionway smashing the wood and iron shelter to flinders.

Barely escaping with their lives, the four men sheltering within were suddenly exposed to the full force of the frequent squalls and surf. Faced with no other option Captain Harrower ordered James Revell, George Carder, George Blackman aloft, to join the others already hanging on for dear life in the mizzen rigging. The predawn light revealed a day of cold, stormy weather with heavy passing showers. The wind was blowing between 20 and 30 knots with frequent strong gusts of up to 50 knots from the west-southwest. Those aloft could just see a small crowd of onlookers gathering upon the shore. The bereft sailors could but hope that the people safe upon the sand had already notified the port authorities and that help would soon arrive.

Just after 7:00 am James Revell decided that it was time to leave, Captain Harrower agreed and they, along with George Blackman descended to the poop. Revell was determined to get ashore and said as much to his Captain. The two men began to disrobe, Revell helping Russell Harrower off with his coat, but then he had a change of heart seeing that there was little hope of surviving in the wreckage filled seas. Russell then climbed back aloft leaving James alone on the windswept poop. Stripped down to his undergarments the doughty sailor leapt from the rail and within moments was carried away aft by the surging water. He quickly grabbed a small piece of flotsam and swallowing a great deal of water still managed to paddle his way through the riotous surf. After a struggle of nearly twenty minutes, the exhausted seaman was hauled from the water by the hands of several brave rescuers who were willing to risk their own lives to bring the man ashore. More dead than alive, Revell was quickly taken up the beach where he was attended to. Wrapped in a blanket the plucky sailor was whisked away to the welcoming warmth of the Seaview Hotel, a half-mile down the coast at Port Willunga.

Perhaps encouraged by Revell's efforts John Hazeland rallied those in the fo'c'sle to attempt a launch of the ship's lifeboat located abaft the half-deck house.

Star of Greece

Port Willunga residents do their best to rescue the sailors trapped upon the Star of Greece, circa 1888.
South Australian Register.

Star of Greece

Despite the seas constantly flooding the deck a number of the stronger sailors sallied forth to prepare the boat for launch. The boat needed to be lifted bodily from the chocks. The sailors had just removed the lashings when a roller shipped aboard and lifted the boat clear. The now wayward craft slid to leeward and was smashed against the bits and stanchions, her sides being crushed and splintered. The seamen attempting to launch the boat barely had time to leap clear. Having failed to launch the deck boat several of the men led by John Hazeland staggered forward to swing one of the aft boats out on the davits. Conditions on deck were far too treacherous, with seas threatening to sweep the men overboard with every step, and thus defeated, the men beat a hasty retreat to the fo'c'sle.

Whilst forward the waterlogged sailors could not fail to observe that the ship was beginning to break apart plank by plank, warping and bending under the weight of water and swelling grain in the aft hold. The men in the mizzen rigging also noticed this predicament with increasing horror. Swollen bags of wheat were now floating freely out of the hatchway and deck planking was snapping and splintering ominously as the iron deck plates buckled beneath. Carl Claeson was determined to make his way forward and called upon those aloft to follow. The agile young Swede shimmied up the ratline and leapt across to the mizzen forestay, following close behind at a much slower pace came Robert Muir and John McVicars. Sliding as fast as their fear would let them the two landsmen moved hand over hand down the stay, safely making the top of the mid-deck house. Timing their run between waves the trio stumbled quickly towards the fo'c'sle and the illusion of safety it offered.

Seeing the three men arrive forward unscathed a further three decided to chance their luck. The carpenter Robert Donald, and two apprentices Alf Prior, Charlie Commerford quickly followed Claeson's example making their way down the forestay and onto the deckhouse roof. From there they had to work to avoid being swept overboard. They hurried from the shelter of one deckhouse to the next before dashing the port fo'c'sle hatchway. The crew quarters were now an overcrowded refuge filled with the stink of fear and vomit as the ship slowly began to break up around the trapped sailors. Their

Star of Greece

only hope was that a steamer or rocket gear would arrive soon to send a line out to the ship before it was too late.

By 9:00 am those aboard the stricken vessel could see a large crowd gathered less than 300 yards away upon the beach. They were attempting to shout out to the men aboard the Star of Greece but the roaring of the wind and waves made it impossible. Having failed to launch any boats Hazeland was determined to save his fellow crew members. With Captain Harrower trapped aft, he felt it was his duty to try to carry a line ashore. A strong but light hemp rope was tied about his waste and with the help of several stout sailors, the first mate was carefully lowered over the side, abaft the fo'c'sle ladder. Hazeland struck out for the shore but soon found himself being swept back to the ship having got within thirty yards of the beach. He was eventually washed back to the side of the clipper and had to be hauled aboard, exhausted but not without hope. In his breathless state, he had not yet found the energy to inform the rest of those within the fo'c'sle of a possible solution to their problem. Once aboard John discovered that the forward wooden bulkheads that divided the fo'c'sle off from the main deck were stove in and water was continuously sweeping in leaving the frightened men trapped up to their wastes in freezing water. No one had eaten or had anything to drink since midnight and those still aboard the Star of Greece were becoming increasingly desperate.

It soon became apparent that it would now be every man for himself. Two friends, Bill Waugh and Robert Donald, a pair of adventurous Irish sailors were determined to try their luck in getting ashore. Despite Hazeland's impassioned pleas the two men together leapt over the side from abaft the cathead and immediately set out for the shore. The plucky sailors battled the wind and waves valiantly for six or seven minutes, becoming increasingly exhausted with each passing wave. At last, a particularly large comber broke over them and Waugh and Donald disappeared beneath the surface still clutching each other's hand. Horrified despair filled those aboard the Star of Greece as the men's bodies appeared, lifeless in the backwash. The twin corpses were washed aft coming within thirty yards of the poop deck before sinking beneath the waves.

Star of Greece

Wreckage strewn beach at port Willunga during the destruction of the Star of Greece.
The Australasian Sketcher. State Library of Victoria.

Star of Greece

Port Willunga Jetty during a westerly gale, circa 1900.
State Library of South Australia.

Star of Greece

Their bodies were drawn down by the current and became trapped against the hull of the clipper somewhere near the rudder. Soon after the two men disappeared a disaster of much greater proportions struck the Star of Greece. With the hull of the clipper canted increasingly to seaward vast amounts of seawater now filled the aft hold. Deep within the irresistible pressure of swollen wheat against the rivets holding the strakes in place quickly began to tell with catastrophic finality. Those perched in the mizzen rigging felt a violent lurch accompanied by the squeals of rent iron plates and the soul-wrenching cracks of deck planking. The stern section of the Star of Greece just abaft the half-deck house shifted suddenly to port as it was torn away from the rest of the ship. The jagged shattering of the plates allowed water to rush into the main hold quickly washing thousands of bags of wheat out into the roiling surf.

The stern canted wildly as the mizzen mast toppled over shoreward dumping Russell Harrower, George Blackman, George Carder and Andrew Blair into the raging sea. The four men stood almost no chance as the iron mizzen mast and cro'jack came crashing down upon them. Knocked unconscious by either wreckage or the fall, all suffered head injuries to some degree and were soon drowned. In the swirling waters trapped beneath the collapsed stern, the bodies of Robert Donald and William Waugh were crushed. Donald's body was pulverised whilst Bill Waugh's corpse was flensed by the shifting weight of the hull upon it. Whilst the meatier parts of the body were mashed and mangled the poor man's hide, scalp and skin from his legs and one arm were ejected out into the surf to drift about becoming food for fish.

Shocked and horrified by the sudden deaths the remaining sailors were determined to stay put to until the arrival of help from a steamer or those gathered on the beach. By 10:00 am an even larger crowd of ghoulish sightseers and would-be rescuers had gathered along the cliff tops and shoreline. Those aboard the Star of Greece could see those on the beach opposite attempting to get a flimsy raft out into the surf to float a line to the ship. Unfortunately, a rip in the lee of the wreck prevented any floating object from getting more than 100 yards from the shore before being pushed out sideways and thence back to the beach.

Harbourmasters Cottage, Port Willunga.
Onkaparinga City Library Archives.

Star of Greece

Pier Hotel (Uncle Tom's cabin), Port Willunga, circa 1895.
Noarlunga City Archives.

Star of Greece

Later in the morning one of the surviving men observed a cabin door from the now missing mid-deck house being held up by several figures onshore. Upon it he could just make out the words *"Float a line!"*, and it was then that the sailors were determined to rescue themselves.

The seamen set to with a will cutting several lengths of rope from the fore rigging to which were lashed three to four empty sea chests. Each attempt brought brief moments of hope yet within half an hour the little craft had floated halfway out towards the beach before being swept back to the side of the ship and dashed to pieces. The men were now beginning to lose hope and many despaired of ever being rescued. They began to wonder why no one with rocket gear had arrived or why those ashore had not attempted to launch one of the many large fishing boats they could see up by the jetty. None aboard could know that there was no longer a seaworthy life-boat or rocket gear stationed at Port Willunga, nor that authorities back in Adelaide were vacillating over a decision to send help.

By midday with the weather continuing cold and squally, the men aboard the Star of Greece had not had anything to eat or drink for more than twelve hours. A few sailors were near catatonic with fear and exhaustion as they sought shelter beneath the fo'c'sle head. A lookout lashed to the topgallant rail could see a large blackboard with something written upon it in large white letters. Unfortunately for the men, the message could not be read, John Hazeland's telescope having been lost sometime during the morning. The message had been chalked onto the board by Constable Tom Tuohy the local trooper from Willunga who was attempting to co-ordinate the shore-based rescue efforts. He scrawled a large yet simple message to let the stranded sailors know that help was coming. Even though those aboard the Star of Greece were unable to make out the message they chose to stay put for the time being. An exhausted Gus Carlson, unable to take anymore trapped within the fo'c'sle climbed wearily upon the bow of the clipper and lashed himself to the rail to wait out the time until rescuers arrived.

First Officer John Hazeland, circa 1888.
R .Tuohy Collection, Willunga.

Star of Greece

Back down below within the flooded confines of the overcrowded crew quarters, Hazeland could see that his men were losing hope. He became determined to lead by example and began looking for a way to save the more than twenty men still alive. The first mate worked hard to raise the sailor's spirits yet, in the end, he decided that the only way for him to save them was to set an example for the miserable wretches to follow. Thus at 1:00 pm he and three young sailors stood upon the fo'c'sle head whilst others watched from lee portholes. Hazeland stripped down to his underwear and clambered out to the end of the bowsprit with the greatest of care. The three boys James Johnstone, Frank Kearney and Henry Corke watched and waited, calling encouragement to Hazeland as he prepared to jump into the raging sea. The first officer took his time watching for when a particularly large wave rolled across the front of the ship.

With little hesitation, he leapt for his life plunging into the icy water. Riding upon the front of a comber as it broke John was carried for more than a dozen yards by the force of the water before the crest broke sending him deep beneath the surface. Pushing up from below the struggling sailor turned over upon his back and road the inward current to within a few yards of the beach. The next thing he knew several strong hands were lifting and dragging his frozen and waterlogged body from the surf. The mate had been in the water for just over five minutes. Quickly wrapped in a warm blanket Hazeland was whisked off the beach and into the welcoming warmth of the Seaview Hotel, there to join the much recovered James Revell.

Seeing their first officer safely ashore Johnstone, Kearney and Corke leapt one after the other from the base of the bowsprit, a grave error that saw them being swirled around in the backwash between wreck and beach. The boy's struggles became even more desperate as they clutched onto anything that would help to keep them from going under. The trio bobbed around within the wreckage strewn maelstrom for almost fifteen minutes before being swept back towards the side of the ship. Those on deck attempted to throw them a line and eventually an exhausted and distraught Henry Corke was hauled back aboard and tended to. Frank and James meanwhile were once again caught in the cyclic movement of the churning white waters close to shore.

Star of Greece

This time they were much more fortunate. Constable Tuohy and another man raced into the surf zone with a rope tied to their wastes and pulled the bedraggled apprentices to safety. There was much-heated discussion amongst those left aboard, some wanted to wait for rescue whilst the more desperate were determined to get ashore anyway they could. One such was Alf Organ, who by now was gripped by uncontrollable hysteria. He and Irvine leapt from abaft the cathead and upon popping back to the surface immediately thrashed their way to a fallen spar. The seamen were making good progress when an enormous, wreckage filled wave rolled over the top of the men knocking both out cold. Neither surfaced again for several minutes and in the end, only Alfred Organ's body washed ashore.

By 2:30 pm the those left aboard the Star of Greece were; Charlie Commerford, Gus Carlson, Ed McBarnett, Alf Prior, Jack Airzee, John Gatis, Peter De Smet, Carl Claeson, Henry Corke, Wilhelm Oerschmidt, David Bruce, Bill Wiles, Eduard Bluhm, Robert Muir, and John McVicars. Each was becoming more resigned to his fate as the arrival of help looked less and less likely. The sailors were now all but leaderless, the captain, second mate and boatswain were dead, the first mate had abandoned them and they had little faith in third mate Charlie Commerford, who was still just a fourth-year apprentice not yet out of his time.

John Gatis decided he could wait no longer. Stripping down to a shirt and trousers he leapt from the fo'c'sle head and in the water grabbed a piece of wreckage. He then spent the next thirty minutes struggling to gain the shore. Already cold and hungry and suffering the effects of exposure the distraught sailor was finally hauled from the surf in an unconscious state. Following the medical thinking of the time, his would-be saviours stripped off his shirt and began rubbing alcohol across his chest and used various methods in an attempt to reanimate the comatose sailor. They worked on him desperately for almost two hours before the man rallied briefly. His eyes opened, and he attempted to sit up, then just as quickly suffered a massive heart attack and died on the beach in Maria Bowering's arms.

Apprentice Frank Kearney, circa 1888.
Richard Touhy Collection.

Star of Greece

Just after 2:00 pm the last of the bulwarks sheltering those within the fo'c'sle failed, leaving the men exposed to the full fury of the wind and waves. At times the miserable sailors were up to their necks in freezing seawater as it sloshed around their bunks. This untenable situation convinced De Smet and Blair to attempt the swim to shore through the tumult of wreckage filled surf. Taking a terrifying leap of faith, the desperate sailors lashed themselves to a broken spar. They struggled in vain for between five and twenty minutes before succumbing to injury and exhaustion. Both were drowned. Not everyone was so unlucky, an observant pair of seamen, David Bruce and Eduard Bluhm, both strong swimmers, noticed the dangers of using wreckage to make the beach safely. Eschewing any form of flotation aid the two men set forth from the base of the bowsprit. Following the example set by John Hazeland, they swam out to the edge of the rip and used the wave action to push them ashore. Both arrived safely within ten minutes cold and tired, but also alive. They too were pulled to safety by Port Willunga locals and bustled off to the Seaview Hotel where they joined the fortunate few already ashore.

Perhaps feeling that he should show leadership, Charlie Commerford along with Willie Oerschmidt attempted the dangerous swim to the beach. By now the weather had begun to abate somewhat, and patches of blue sky appeared amongst the scudding grey clouds. The winds were as strong as ever and the seas as high. Still, the desperate duo was willing to swim the gauntlet. Of the two brave sailors, only the third mate managed to reach the welcoming arms of his rescuers. The remaining apprentices, Ed McBarnett and Alf Prior took a more methodical approach to their self-rescue efforts. Breaking into the side-lamp casing young Prior stripped off and covered himself in lamp oil. Once fully oiled up he climbed the rail and jumped into the surf. The oil coming off his body did its job smoothing out the water immediately around him. The clever young apprentice was soon struggling through the foaming shallows as several rescuers rushed in to help. Ed McBarnett was followed ashore soon after by Alf Prior. He came onto the beach with a bunk board under each arm, and though unable to swim had managed to stay afloat long enough for rescuers to pull him from the water. It had taken him all of eight minutes to reach the beach.

Star of Greece

**Desperation drives last survivors of the Star of Greece to attempt
to reach the shore.**
Willunga Old Courthouse Museum Collection.

Star of Greece

Seaview Hotel, Port Willunga circa 1885.
State Library of South Australia.

Star of Greece

The sun was low in the sky and the wind again picking up when four sailors emerged on deck carrying a crude triangular raft made from oars and bunk boards lashed together. They carefully lowered their flimsy vessel over the side and leapt blindly into the surf by the side of the shattered clipper. One of those aboard was John McVicars who by now was regretting ever having boarded the Star of Greece to evade the police. The raft was never going to stand up to the battering it received in the pounding surf. After being churned around for some time the flimsy vessel finally broke up dumping its hapless occupants into the water. None of them it appeared could swim, and of the four none reappeared once they were swallowed by the waves. There were now just four men left aboard the stricken vessel; Carl Claeson, Jack Airzee, Henry Corke and the veteran seaman Gus Carlson. The disaster was now entering the end game as pallid white-gold rays of the setting sun shone briefly upon the horizon.

Claeson decided that there would now be no rescue forthcoming and he would have to risk all in a bid to reach safety. He hauled out a bench from the fo'c'sle and knocked the feet from it with the help of his friend Jack Airzee. They thought the plank, 12 feet long and 1 foot, wide was the perfect vessel upon which to float to shore. Jack pleaded with Carl to be allowed to accompany him for he was not a strong swimmer and did not want to be left behind. Using rope cut from a halyard, Carl then made a loop at each end of the plank and together the two young men carefully lowered the plank over the side. Tying it off to the rail they then stripped down to their underwear and waited as a series of large waves broke over the hull. Using another line they gingerly lowered themselves over the side. They then looped the rope at each end over their shoulders so that the plank rested comfortably upon their chests.

With knives in hand Claeson and Airzee cut themselves away to float out into the still heaving surf. The wash drove the two sailors aft towards the crumpled stern of the ship and they feared that the floating wreckage would kill them before they reached safety. It was then that a large breaker rolled right over them and pushed the two desperate men onto the sandy bottom. Claeson's feet became tangled in weed and Airzee was struck on the head by the plank as the waves pummelled them.

Star of Greece

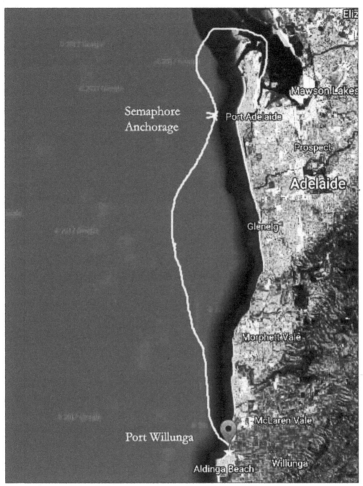

The last voyage of the Star of Greece.

Star of Greece

Carl came up holding his breath whilst an unconscious Jack was held face down in the water by the weight of the plank upon his back. Carl Claeson was unable to help his injured friend and soon passed out from exhaustion and cold. The two sailors slowly drifted ashore and were soon hauled clear of the surge. Once on the beach rescuers discovered that one of the two was still alive and he was wrapped in blankets and taken off to the home of farm-owner Maria Bowering. Jack Airzee, on the other hand, was laid respectfully upon the beach alongside the other five men whose bodies had already been recovered.

There were now but two men left aboard the Star of Greece, Gus Carlson, the aged sailmaker and a Henry Corke who was desperate to get ashore. Divesting himself of all his clothes he dove into the surf from the bowsprit, attempting to emulate the efforts of those who had successfully made it to shore. However, by now, the water was filled with debris from the wreck and though the young lad started strongly for the beach a ferocious comber picked up a spar as he was passing and dumped it upon Henry's skull. The unfortunate soul was struck viciously and knocked out cold, his motionless body floating briefly before disappearing less than one hundred yards from the beach. He was never seen again. With darkness almost upon him, Gus decided that help was never going to come.

Refusing to undress knowing that cold was as great an enemy as the water Gus made his way carefully to the centre of the ship seeking the shelter from the breaking waves in the lee of the gutted remains of the deckhouse. Then with little hesitation, he removed his boots and dropped into the water. Once on the surface, Carlson began the long and desperate swim. He was carried over a large wave and managed to latch onto the sizeable piece of wreckage. Hanging onto it for dear life Gustav battled on but was soon carried back and forth through the eddying waters from wreck to shore and back again. After almost fifteen minutes the aged sailmaker had been carried to the shattered stern of the clipper. Exposure and exhaustion now began to take their toll as the Sailmaker's strength began to fail.

Star of Greece

Gull Rock, Blanche Point where Captain Harrower's body washed ashore.
State Library of South Australia.

Star of Greece

Firing the rocket gear at Port Willunga, circa 1915. The essential lifesaving gear arrived too late to be of any use at all.
State Library of South Australia.

Star of Greece

Turning over onto his back he clung on to the piece of wood that was keeping him alive. Gus drifted out past the stern when a mountain of white green water smashed down upon him driving the nearly unconscious man beneath the surface. When he, at last, emerged it was face down, unmoving. Gus Carlson's lifeless body soon after washed ashore, his struggle for life in vain. His saddened would-be rescuers could do naught more than carry his much battered and bruised body to lie alongside his deceased colleagues high up on the sand.

In a cruel twist of fate, the much-delayed rocket gear finally arrived from Normanville, the heavy rescue apparatus having to be manhandled across the dunes. Those carrying it crested the sandhills just as Carlson's corpse was being hauled from the surf. It was a forlorn sight that greeted the men as they joined the distraught and horrified locals around a huge bonfire that had been lit to warm those rescued. The beach was awash with shattered timbers, the personal belongings of those once aboard, the waterlogged carcases of pigs and chickens, ship fittings, and thousands of split hessian bags of wheat. The wreck itself sat torn asunder in the rolling surf, a shattered reminder of the day's tragic events.

By now the survivors were being well cared for by Port Willunga locals. John Hazeland, Carl Claeson, Frank Kearney and Ed McBarnett had been lead to the home of Maria Bowering, not far from the scene of the wreck whilst Charlie Commerford, Alf Prior, James Johnstone, David Bruce, Eduard Bluhm, and James Revell were warming themselves by the fire at the Seaview Hotel. As the evening wore on six bodies were reverently carried up off the beach and taken by horse-drawn dray to the Seaview Hotel. This was a large two-story stone building owned by William Kimber who made his place of business available for the use by Thomas Tuohy until an inquest and funeral could be held. The bodies were stored in a small stone cottage at the back of the pub. The dead were joined later by the flensed hide of Bill Waugh; a four-foot square piece of back-flesh and portions of the ear, face and scalp with a shock of bright red hair still attached.

Saturday morning dawned cold and clear with police and volunteers scouring the beach for bodies and possible survivors. Lead by Lance

Star of Greece

Corporal Nalty the men found three more corpses which were quickly conveyed back to the cottage in time for the inquest that was to be held later in the afternoon. The Star of Greece lay in the surf a shattered hulk devoid of life. Her stern section lay torn off angled towards the shore and two lifeboats, one off the chocks, the other on the davits, lay perfectly serviceable aboard. A group of sailors from the steamer Yalata, which had arrived that morning, boarded the wreck. The men carefully inspected the ship and found no one left aboard. With nothing more to be done the dejected sailors returned to the Yalata which was then sent back to Port Adelaide having dropped off fresh clothing for the survivors.

The inquest into the deaths of those men recovered was held at the Seaview Hotel on the afternoon of July 14th. John Hazeland was asked to identify those men he knew, namely George Blackman, William Waugh, Robert Donald and the young Cabin Boy George Carder. Not being acquainted with the eleven seamen and three stowaways he misidentified one body and was unable to positively identify the rest. It was left to others to put names to the faces of the deceased. In addition to those positively named by the mate, the other six given their identities back were John Airzee, Henry Corke, Alfred Organ, Robert Muir, Andrew Blair, and Gustav Carlson. The coroner in consultation with a jury of Willunga locals concluded that the men died as a result of having suffered injury and death by drowning in their attempts to reach the shore from the wreck of the Star of Greece. That evening eight plain wooden coffins (the only ones available) arrived on the back of a wagon and the remains of the ten men were placed within.

The funeral was held on Sunday afternoon in the grounds of the Wesleyan Cemetery. A huge crowd gathered, the largest ever seen for a funeral in the district. In attendance were the many people involved in the rescue of the ten survivors and officials from the police, members of the Maritime Board as well as a sea of curious onlookers. The most poignant of attendees were the men and boys who had lived through the terror of the wreck and were just thankful to be alive. Most would later suffer survivor's guilt and the trauma that came with it. The coffins were interred in a mass grave and by late afternoon many people had begun to drift off home

226

Star of Greece

The wreck of the Star of Greece the day after!
State Library of South Australia.

Star of Greece

Constable Thomas Touhy wearing the medal awarded to him for his work in rescuing sailors from the Star of Greece.
Willunga National Trust Museum.

Star of Greece

The surviving Star of Greece crew were conveyed back to Port Willunga to await transport to Adelaide. They had been offered a berth aboard a steamer sailing to Port Adelaide but each had roundly refused to set foot aboard another ship so soon after the wreck and elected to return to Adelaide by coach. The exception to this was John Hazeland, he jumped at the opportunity to clear out from Port Willunga and arrived in Adelaide late on Sunday evening. From the Port, he was taken to the private residence of friends to await the inevitable summons to the upcoming Marine Board enquiry.

The hunt for bodies continued and police found three more unfortunate souls over the next ten days. The German sailor Wilhelm Oerschmidt was found wedged amongst boulders near Gull Rock on July 15th, not long after the funeral had taken place. His body was taken to the little cottage behind the Seaview Hotel to await burial. He was laid to rest in a quiet service within the grounds of St Anne's cemetery on the outskirts of Aldinga village. Captain Henry Russell Harrower was discovered by Trooper Thomas Tuohy who had great difficulty in lifting the body from the base of Gull Rock on July 16th. He was interred alongside his men in the mass grave the next day. His brother-in-law James Bishop was in attendance as were many locals who had participated in the rescue attempts on July 13th, a grief-stricken Alice stayed in Adelaide, too overcome with melancholy to make the arduous journey south. Police kept up the vigil for another week searching as far north as Port Noarlunga.

The last corpse to wash ashore was found on July 23rd also near Gull Rock. It belonged to George Irvine, it was obvious from the dead man's ghastly appearance that small sharks had been feasting on the body. The much-decomposed individual was quietly buried alongside his shipmate in St Anne's that afternoon. There were few people in attendance that day the community still reeling from the shock of the horrendous deaths of eighteen men and boys. The bodies of William Parker, Peter De Smet, John McVicars, John Gatis, and William James Miles were never recovered though parts of unidentifiable bodies continued to wash ashore between Port Willunga and Maslin's Beach for many weeks to come.

The original obelisk erected by the Aldinga community in
September 1889.

Star of Greece

Monument Inscription, circa 1889.

The local community of Port Willunga/Aldinga raise the money for the monument via subscription. Thus by the time, it was erected the record of just who was interred beneath had been lost. The names and graves of the two sailors buried at St Anne's cemetery across the road were all but lost and in a time forgotten.

"Oh Hear Us When We Cry To Thee
For Those In Peril On the Sea"

Star of Greece

Figurehead salvaged from the Star of Greece, now on display at South Australian Maritime Museum.
Photograph by author.

Compass housing from the poop of Star of Greece.
On display at the South Australian Maritime Museum.
Photograph by author.

Star of Greece

**Ships fire bucket salvaged from the wreck and presented to
Constable Thomas Tuohy along with the other 'E' marked bucket
as a souvenir of the day of the wreck.**
On display at the Willunga National Trust Museum.
Photograph by author.

Starboard navigation lamp salvaged from the wreck.
On display at the Willunga National Trust Museum.
Photograph by author.

Star of Greece

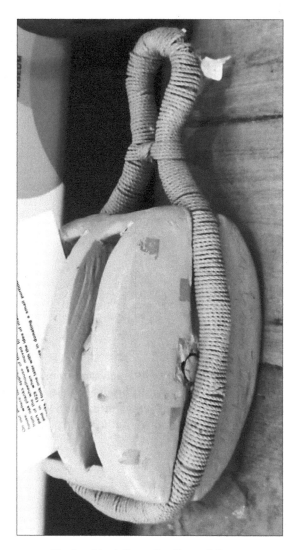

Rigging block from the Star of Greece.
On display at Willunga National Trust Museum.
Photograph by author.

Star of Greece

A Trotman's anchor, one of several salvaged from the Star of Greece (with Dick Jagoe). **It now sits proudly on display in front of the Mariners Memorial at Semaphore, South Australia.**
Photograph by author.

Star of Greece

**Wooden seatback bearing the JP Corry crest and a brass belaying pin salvaged from the wreck.
On display at the Willunga National Trust Museum.** Photograph by author.

Captain Harrower's dining set, salvaged from the Star of Greece.
Photo by Hans Ehmann.

Star of Greece

The bell of the Star of Greece salvaged from the wreck and now in private hands.
State Library of South Australia.

Star of Greece

The shattered remains of the wreck were sold by auction at the Port Adelaide offices of Elder & Co'. The cargo was insured for £12500 by a variety of underwriters yet the ship itself was uninsured, Corry's having carried the risk due to sharp increases in insurance rates for sailing ships. There was little real interest in the wreck, the hull being sold for scrap for £105 to well-known Port Adelaide Ship's Chandler William Russell. He also paid another £5 for all of the ship's fixtures, two boats, cable, anchors, masts, spars, sails, rope and rigging, capstans, windlass, donkey engine parts and stores. The 1840 tons of wheat in 16002 bags was knocked down to the bargain price of £21 to local miller Mr W.J. Tonkin who planned to salvage the remaining wheat for stock feed, malting and milling. William Russell's men spent the next month stripping the Star of Greece of everything worth taking. The surviving sails, boatswains' stores, royal and topgallant masts, spars, boats, cable, rigging, the fo'c'sle capstan, top of the windlass and all of the anchors were eventually shipped to Port Adelaide aboard the steamer Cowry.

It was found when the port bower anchor was raised that the cable had fouled around the stock thus preventing it from ever gaining purchase at the time of the wreck. This fact was not known to the Marine Board at the inquest. Much of the wheat that washed ashore was dried in kilns for malting by workers from Tonkin's mill. By the middle of August, more than 1000 bags had been recovered from the forward hold and being almost completely dry turned a tidy profit for the canny miller. By the end of September, nearly 4000 bags of wheat in fair to good condition would be removed from the forward hold. Most of the lesser fittings from the wreck were auctioned on the beach, amongst the items put up for sale by William Russell were the ships figurehead, fo'c'sle bell and fire buckets.

The nine remaining survivors from the Star of Greece were placed aboard a horse-drawn coach supplied by Hill & Co', in the wee small hours of Monday morning for the long and bumpy trip to Port Adelaide. They arrived there just before lunch the same day accompanied by Police Detective McKinnon and the Seaman's Runner Mr Crowe. The wagon was met by a large crowd on the Lipson Street corner who followed it onto the Prince Albert Sailors Home. As the men stepped out onto the pavement, the gathered crowd gave three hearty cheers. The sailors were warmly greeted by the mayor of Port Adelaide, Mr John Cleave, and the head of the Seaman's Mission, Mr Emanuel Hounslow. Mr Cleave gave a speech in honour of the survivors offering them unstinting support and sympathy over their recent plight.

"It is with very great pleasure I welcome you to Port Adelaide, although it is under the most trying circumstances. We need not say anything today about the very great sufferings you have endured during the last few days. I can only say that the people of South Australia will, to a man, try to atone for the very great negligence we have shown in your case. On Saturday night there was a crowded meeting in the town hall, and your mates felt and spoke very strongly of the negligence and lack of judgement shown in trying to rescue you from a watery grave. It was determined that something should be done for those that have been spared...Nothing we can do will be adequate to atone for the loss of your comrades, whose lives are lost, but whose memories we can recall. I can promise you that you will have the sympathy of all the people in the colony. I will see that provisions are made for your bodily comforts, for your proper clothing, and that you receive the kindest attention at the home."
South Australian Weekly Chronicle, Saturday 21 July 1888, page 6

The nine men and boys were then taken around by Mr Hounslow and his assistant Mr Hill to many of the local shops so that they could buy what items of clothing and kit that felt they needed to replace that which was lost. All of the sailors were exhausted and much traumatised by their ordeal and it would be many days before they felt well enough to face the outside world once more. The castaways were given little opportunity to rest. The local community began to rally around them organising fundraisers aplenty to see the men clear financially and to help raise money to send to Britain to provide some form of financial recompense for the parents, widows and orphans back home. All the seamen required ongoing medical attention for their wounds both physical and psychological and yet they were in much demand at the various concerts, picnics and galas, held in their honour.

Star of Greece

Mission to Seamen, Todd Street, Port Adelaide, circa 1910.
Gordon Walker, photographer, State Library of South Australia.

The Customs House at Port Adelaide, 1894.
Searcy Collection, State Library of South Australia.

Star of Greece

Of the ten survivors of the Star of Greece, the four Able Seamen; Edward Bluhm, James Revell, David Bruce and Carl Claeson were accommodated at the Seaman's Home and had their needs taken care of by the Mission to Seamen represented by the affable and tireless Emanuel Hounslow. The four apprentices; Frank Kearney, Alfred Prior, James Johnstone and Edward McBarnett along with Charlie Commerford whilst also staying at the home had their financial and material needs covered by Corry's agents in Port Adelaide who then forwarded the account to London for settling. Of the ten survivors, only John Hazeland kept himself aloof from the goings-on in the aftermath of the wreck. He remained in self-imposed isolation staying with friends until the day of the Marine Board enquiry which was to be held the following Friday, July 20th 1888.

The Marine Board preliminary enquiry into the loss of the Star of Greece began promptly at 9:00 am. Held within the chambers of the Port Adelaide Customs House, it was presided over by the Chairman Mr John Formby and Marine Wardens; Legoe, Lyndsay, McCoy, Muecke and Smith, all were experienced ships masters in their own right. Also present was the Secretary Mr Thomas N Stephens who was responsible for the administration and proper running of the investigation. All these men were well aware of the public and political anger being directed at them for their perceived dithering and quibbling whilst the Star of Greece's crew were dying at Port Willunga.

Star of Greece

The Blame Game

"There is never a wreck in Saint Vincent Gulf but as the result of willful and gross carelessness."

John Formby, 1888.

The members of the Marine Board realised that they had not exactly covered themselves in glory that day and were understandably looking to deflect attention away from their apparent callousness in the face of such tragedy. What was less well known by those in the gallery was that the Marine Board had had its budget severely trimmed in previous years by different penny-pinching colonial treasurers. They were also under strict instructions that permission for any unusual or extraordinary expense requirements needed the tacit approval of the Treasurer or his Secretary. Unfortunately, the Port Adelaide lifeboat, and the Port Willunga rocket gear, had not been used in decades so had been allowed to deteriorate. The required funds had instead been spent to maintain the more recently utilised Port MacDonald lifeboat and the rocket gear of the Cape Borda and Cape Willoughby lighthouses on Kangaroo Island. The Marine Board chairman John Formby was under constant pressure from Treasury to rein in costs and to work within a very strict budget. This was despite having over 2000 miles of coastline to protect. Thus, it was under such scrutiny that the members of the Board met on Monday, July 20th to determine fault for the wreck of the Star of Greece and subsequent loss of so many lives. The Colonial Government was also looking on with a vested interest in the eventual findings of the Board's investigation.

The Chairman Mr Formby called several witnesses to explain the events leading up to the loss of the Star of Greece, the conduct and character of Captain Harrower and the catastrophic loss of life after the vessel's stranding. The terms of reference of the inquiry made almost no mention of delays in the arrival of the rocket gear or of the penny-pinching decisions made by the Board that meant that no genuine rescue attempt was even considered by the crew of the HMS Defender until it was too late. The first witness called was pilot

Star of Greece

William Boord, he detailed events leading up to the Star of Greece leaving Semaphore.

Next to give testimony was John Hazeland who was examined for almost two hours by members of the committee about all aspects of the disaster. The Board called into question his decision to abandon ship before the rest of the crew were safely ashore. This assertion was later disputed by Carl Claeson who stated that Hazeland's actions had inspired at least some of the crew to follow his lead. This was not a view supported by all survivors. The committee then adjourned for lunch. In the afternoon Mr James E Dempster, a senior member of the firm Dempster, Webb, & Dempster, solicitors, of Port Adelaide, speaking on behalf of the ship's agents and owners, asked to call witnesses of his own.

The first to take the stand was apprentice James Johnstone who was questioned at length by the Wardens. He was followed by Alfred Prior, David Bruce and third officer Charles Commerford, whom all gave their versions of events aboard the clipper up to the time she struck. John Hazeland was recalled later in the afternoon for further cross-examination, he was followed by perhaps the most dubious character of the lot James Revell. The man had previously stated to reporters that Captain Harrower had been drunk in charge of the vessel and that was why it had gone ashore. A former carpenter of the Ipswich shipyards, Revell had abandoned his wife and young daughter back in England and was a known to local police as a heavy drinker and trouble maker. Despite this, he was a witness to events and the first man ashore. Having been given corrective treatment by the rest of the survivors at the Seaman's Home he recanted his earlier statements in the media about Henry Harrower. However, by then the damage had been done and it gave the Chairman Formby a convenient hook upon which to hang the Board of Enquiry's eventual findings.

The Marine Board members adjourned for a short time just before evening and there was perhaps much-heated discussion amongst men about where to lay the blame. It was obvious to all present that the Marine Board held a great deal of culpability for the failure to rescue any of those aboard the doomed vessel. Yet there was also the public and political anger directed against the Board members to consider.

Star of Greece

In the end, the committee Chairman, Mr John Formby overruled all objections and soon after delivered his, and by default, the Marine Boards findings. He read out the verdict to the waiting crowd.

"That the Board having carefully inquired into the circumstances attending the loss of the Star of Greece find as follows; A manifest want of skill was shown in the management of the vessel from the time of starting from the anchorage, and the time required to approach Backstairs Passage by daylight was miscalculated. Laying the ship to from 9 pm until 2 am, on a lee shore with the main topsail aback, and a strong gale blowing was fatal. The deep-sea lead should have been used, more especially as the ship was drifting. The master having lost his life and no blame being preferred against the mate, the Board do not propose to refer the case to a Court of Marine Enquiry. There is nothing to warrant there having been drunkenness on board."
South Australian Register, Saturday 21 July 1888, page 5

The findings were a whitewash, impugning the reputation and calling into question the competence of Captain Henry Russell Harrower whilst abrogating all the responsibility being laid at the feet of the Marine Board. The findings outraged many and satisfied no one. The Board of Enquiry had ignored the fact that no storm warning had been officially issued on the day of sailing. They made no mention of the fact that land had been sighted by one of the apprentices at 1:30 am, more than an hour before the ship struck but William Waugh, the officer on watch at the time had ignored the warning and failed to inform Captain Harrower of the sighting. Pilot Alexander Boord failed to mention or was warned not to talk about the discussion that he, Captain Harrower and first officer Hazeland had about leaving when they did. John Hazeland had insisted that they depart as the ship was ready with all sails bent and all clearances having been gained, this despite Captain Harrower being in two minds about departing in the face of the approaching cold front. The Board also did not know of the fouled anchor, a fact quickly established by the salvors who failed to make any mention of it to the Board at the time. The Marine Board laid the fault for the loss at the feet of the deceased Captain Harrower. Knowing that a Court of Enquiry would reveal

Star of Greece

more facts than they wanted to be made known, John Formby ended proceedings then and there.

Public and political reaction to the Marine Board's report was scathing and soon there was enough pressure put upon the government by influential members of the colonial establishment for something more to be done. In response, the government chose to convene a Select Committee charged with investigating the reasons for the tragic loss of life attached to the wreck of the Star of Greece. The first hearing was held on July 23rd and the crew of the Star of Greece were excused from giving further evidence on July 28th, yet proceedings continued until August 8th. There was much debate about the government's ultimate culpability for the debacle that was the bungled rescue effort on July 14th 1888. Only two survivors from the Star of Greece were called on to give evidence, John Hazeland and Carl Claeson. An odd choice considering that the other eight men may well have given a less glowing account of the events than did the two chosen to appear.

The Select Committee was not out to crucify the remaining members of the ship's crew. Instead, they were looking to abrogate any responsibility that may have been laid at their feet by the papers, prominent members of the public, members of the Marine Board and a rather vociferous Parliamentary Opposition. The Select Committee Report was tabled and accepted on August 14th 1888. The findings cleared the Treasurer and Government of all blame for the botched rescue and laid it squarely at the feet of the Marine Board Chairman, John Formby and its long-time Secretary Thomas Stephens. To say that the Opposition and even members of the Select Committee were incredulous would be a gross understatement.

In summary, the Select Committee in its report found that;

i. Members of the Marine Board were not required to consult with the Treasurer when seeking to distribute funds in case of an emergency.
ii. The Board was fully responsible for the adequate distribution and maintenance of all essential rescue equipment such as rocket apparatus and lifeboats.

Mr John Formby, Secretary of the South Australian Marine Board.
Searcy collection, State Library of South Australia.

Star of Greece

iii. That the Board did all that could have been reasonably expected of it during the Star of Greece emergency.

iv. The members of the Marine Board could not say why the rocket apparatus and lifeboat had not been adequately maintained, nor why the rocket apparatus had been removed from Port Willunga in 1883.

vi. The Secretary of the Marine Board had received notice of the wreck at 9:40 am on July 13th yet did not appear to take any noticeable action in response to the unfolding disaster until between 2 and 3 o'clock in the afternoon when it was decided to consult with the Treasurer about how to proceed.

v. That the Board had been negligent in their maintenance of the lifeboat and rocket apparatus at Port Adelaide.

vi. That the actions of First Officer John Hazeland were to be commended and not censured as many in parliament would have preferred.

The Opposition's reaction within the South Australian House of Assembly was immediate and without equivocation. It very soon came to light that members of the Select Committee had attempted to insert a clause into the final report that in essence stated that no steamer with or without a lifeboat aboard could have made any appreciable difference to the fate of those aboard the Star of Greece other than to compel them to stay with the wreck until the rocket apparatus arrived from Normanville. Secondly that the actions of John Hazeland were not as heroic as they were made out to be, and that more lives would have been saved if the first officer had stayed with the men to keep order until help arrived. The clause for insertion moved by the Attorney General Mr Kingston MP stated that;

"We cannot join in any commendation of the actions of the Chief Officer...There was certainly no necessity for their leaving at the time the mate swam ashore...the presence of a steamer was desirable...as it would have encouraged a number of those who were drowned in the attempt to reach the shore to remain on the wreck. So that we think that the action of the mate in setting the opposite example of leaving the wreck cannot be approved. "

Report of the Select Committee of the House of Assembly into the Wreck of the Star of Greece. Government of South Australia, 14th August 1888.

Star of Greece

The motion was seconded by a Mr Dashwood MP another member of the Select Committee, however, the Chairman, a supporter of the Treasurer, David Bews MP, ruled the clause out of order and overruled the motion. The report was tabled and signed, exonerating the Government of all blame and leaving the actions of the Marine Board squarely in the frame. In the weeks that followed a Bill was introduced into the House of Assembly calling for the abolition of the Marine Board. It managed to make it through to the second reading in both houses before being defeated. Tellingly the board and its members were severely censured. The Opposition was rather less sanguine in their reaction to the Select Committee's findings and attempted to push on with the Marine Board Abolition Bill.

"For a long time, the Treasurer and the Secretary for the Board have carried out the functions of the Board...The Treasurer aught get himself a nautical man to act as Secretary...The enquiry into the loss of the Star of Greece shows that the disorganized state of the Marine Board is greatly due to the want of funds following on the Government's system of economy. At the same time, the Board has proved wanting with regard to the stores in stock"

Mr William K Mattinson MP. South Australian Register, Friday 14 September 1888

The political chicanery and shenanigans of local politicians had little effect upon the survivors from the Star of Greece who were more concerned with getting home to friends and family. Each of the Able Seamen was given his discharge on April 21st and placed firmly into the care of the Port Adelaide Seaman's Home and its runner Henry Crowe. All were still in relatively poor health and it would be some time before they would again be ready to go to sea. One of the first to depart Port Adelaide was Charlie Commerford who having completed his time as an apprentice for Corry's agreed to sign on as an Able Seaman aboard the steamer Guy Mannering. The Star of Greece Relief Fund Committee granted him £12 to pay for his incidental needs prior to departure.

Loaded with the second half of the wheat cargo destined for London, the Guy Mannering departed on August 4th for Manilla to take on a load of sugar for Boston. Commerford's voyage home was not without incident, the steamer having to put into Gibraltar en route to Boston, with a leaking boiler. The New England port was reached safely and the Guy Mannering sailed home arriving at Liverpool on Christmas Eve 1888. Charlie caught the first train home to Ramsgate, stopping in London just long enough to receive his final discharge papers from Corry's offices. He was home for the New Year's celebrations and back within the loving embrace of his family.

Charles Commerford gained his second mate's ticket in early March 1889 and went on to serve aboard the clipper Lalla Rookh in the colonial trade. From the Lalla Rookh, he again served with Corry's as first officer aboard the Star of Germany and later the Star of Bengal. Charles Martin Commerford, at last, became a ships master in his own right almost ten years to the day of the wreck of the Star of Greece.

Also set to travel back to Europe aboard the Guy Mannering were David Bruce and Edward Bluhm. Commerford must have been somewhat taken aback when he applied to sign on as a mate aboard the steamer only to find that one who had recently served under him, David Bruce, was now second officer of the homeward bound steam clipper. Born in the town of Montrose in 1864, the son of a brush maker and fancy wears retailer Bruce had begun his career at sea as an apprentice aboard the four-masted barque, County of Caithness in 1880. He continued aboard her for another year working as an Able Seaman before occasionally fulfilling the role of the third mate, for watchkeeping purposes after the death of the third officer of the Caithness, Duncan McDonald in December of 1884.

In 1886 David spent five months as second mate of the 500-ton barque Ben More having just returned from Calcutta. The Ben More sailed in early April for San Pedro to discharge thence to Puget Sound to take on wheat for Britain.

Star of Greece

SS TARAMUNG. 1281 ton, Iron screw steamship, built by Russell at Cartsdyke, Glasgow, Scotland, 1880.
Allan C. Green 1878-1954 State Library of Victoria.

Star of Greece

David Bruce had other ideas and soon after shipped out as an Able Seaman aboard a wheat-laden ship bound for New York, arriving in May of 1887. He spent time ashore haunting the New York Docks before signing on as second mate aboard the 1600-ton ship Kirkcudbrightshire under the command of Captain Alex Baxter. The Kirkcudbrightshire departed New York carrying general cargo for Melbourne on July 22nd 1887. After a stormy start, the clipper made a good showing to the equator which was crossed on August 28th. Expecting a quick easting run Captain Baxter instead found his ship running into a series of dead muzzlers as light and variable south-easterly breezes caused vexatious delays for much of the run across the Southern Ocean. The eventually gave way to the expected westerlies until bitter gales swept in from the southwest which pushed the Kirkcudbrightshire along at more than 15 knots for more than 24 hours at a stretch. Passing Cape Leeuwin another period of light easterlies prevailed until Port Phillip Heads were at last reached on October 17th, 86 days from Sandy Hook. David Bruce stayed aboard until the ship was unloaded and ready to take on her cargo of wool.

Bidding farewell to Captain Baxter he spent more than a month ashore before taking a berth as an Able Seaman aboard the 1200 ton coastal collier ss Taramung which made regular weekly runs between Newcastle, N.S.W. and Melbourne with 1500 tons of coal aboard. David stayed working aboard the Taramung steaming between the two ports in a routine with little variation. Five months aboard the grimy collier was more than enough for the free-spirited sailor who once again longed for the clean air and waters of a blue water trader. He took his discharge from the Taramung in late May 1887 for a run ashore as a stevedore and longshoreman. Bruce soon got itchy feet and jumped at the chance to again set sail. This time he signed on to the barque Edinburgh. The vessel had set sail from Deal on October 15th 1887 bound for Port Melbourne. Commanded by well-known driver Captain James Barclay the barque dropped anchor in Hobson's Bay on January 16th 1888, after 93 days at sea.

Bruce rejoined the barque for her run to port Pirie, South Australia. The vessel arrived at the dusty outport at the height of summer. Bruce stayed on long enough to see the vessel loaded but when she set sail on March 29th 1888, David Bruce had paid off and was on his way to

Star of Greece

Adelaide to look for work ashore and to visit relatives in the colony. It was here he met up with a fellow Scott, George Irvine a fisherman from the Shetlands who had been working the colliers of Newcastle. The duo was looking to change their current circumstances and after time spent working the docks of Port Adelaide. The pair of hopefuls decided to ship out together. Looking to perhaps find a position as mate aboard the Star of Greece, David Bruce soon discovered the position already filled so took what was offered. The two Scott's were entered into the crew manifest as foremast hands on July 10th 1888.

David Bruce stayed with the Guy Mannering for more than a year, before leaving to take his first mate exams in early 1890. He then joined the barque Glenburn as first mate for a run to New Zealand. He never made the shift to ship's master but still had a successful career that lasted until the end of the Great War. The irrepressible Scott had served at age 50 as an Able Seaman aboard the Boarding Steamer, Carrion as part of the coastal defence fleet from 1914 to 1917. The war did not leave the old shellback unscathed as he lost his son James on the battlefields of France in 1917. Having lived a full and interesting life David Bruce finally sailed on to Fiddler's Green in September 1933, aged 69.

The third survivor to take passage home aboard the Guy Mannering was 24 years old Latvian, Eduard Bluhm. Originally from the port city of Libau, a major Russian trade port and home to several thousand German Jews, Eduard had taken to sea at a young age determined to leave the harsh conditions under which his family lived, including the ever-present threat of pogroms. Having survived the wreck of the Star of Greece Eduard was keen to make the shift to steam and the Guy Mannering was that opportunity. He stayed in steamers for the rest of his career at sea until his untimely death from complications of malarial hepatitis contracted aboard the ss Monrovia as she lay docked in Loanda. Eduard passed away as the ship sailed for Buenos Ayres on August 28th 1895.

Released from further testimony at the Select Committee hearings John Hazeland was free to travel to Sydney to visit family. He had in his charge the four apprentices who together boarded an Orient Line

Star of Greece

steamer bound for Sydney en route to London at the beginning of August. The boys imagined they were in for an easy passage home but upon their arrival in Sydney were quickly disabused of this notion.

Instead of sailing on to England, the boys discovered that their indentures had been transferred to the Star of Italy which had arrived in Sydney on August 12th. Hazeland escorted Frank Kearney, Ed McBarnett, Alf Prior and James Johnstone down to Circular Quay where the famed clipper was berthed to load wool for London. They found themselves joining their fellow apprentices Fred Willis, Sam Smyth, and Sam Hogg, in the now overcrowded half-deck under the tutelage of Captain Michael Cotter.

The Star of Italy had just arrived from Vancouver, with a full load of timber, having made a stormy passage across the Pacific in 81 days. After passing through the Heads she was towed into Watson's Bay to await discharge. She was then hauled across to the Johnston's Bay wharf to unload her cargo before being taken across to Mort's dry-dock for a clean and paint. The four apprentices were ferried out to the ship by a steam launch and taken aboard by John Hazeland who paid his respects to Captain Cotter before taking his leave. Michael Cotter was glad of the extra hands as most of the crew had run upon reaching Sydney. As the youngest and newest members of the afterguard the Star of Greece survivors were yet to find their places within the pecking order of the half-deck. After a month in Johnston's Bay, the Star of Italy was taken around to the west side of Circular Quay to await her cargo of wool.

The ship was loaded and almost ready to leave by October 13th 1888 alongside the wool clippers Derwent, under Captain J. R. Andrews, and Achilles, commanded by Captain Robinson. There was heavy betting amongst those in the shipping community as to which clipper would make London first and there was a friendly wager between the three skippers too. Even though the Star of Italy was the firm favourite and attracted the majority of wagers, Captains Robinson and Andrews were determined to lower her colours.

Star of Greece

Star of Italy, 1644 tons built by Harland & Wolff, 1877.
Allan C. Green, State Library of Victoria.

Star of Greece

The 1900-ton, iron-hulled Derwent was first away departing Sydney on October 16th, the Achilles and Star of Italy with 8000 bales of wool aboard, were scheduled to depart on the 17th at almost the same time. The Achilles with 6200 bales of wool in her hold was away, slipping out through the Sydney Heads, at 1:15 pm, after a nasty dust storm had unleashed the worst of its rust-brown fury upon the Harbour City. The crew of the Star of Italy were making final preparations at the departure buoys when a billowing wall of desert dust rolled across Sydney from the west. Such was its sudden violence that the crew had no time to attach a spring to the mooring cable which soon parted. The heavily laden clipper along with several other vessels drifted ashore, bumping heavily against the harbour rocks. Fortunately for Captain Cotter there was no serious damage done and the Star of Italy was towed off by the steam tug Irresistible that afternoon.

Cotter wisely decided to wait until the squally weather abated somewhat. The ship eventually cleared Sydney Heads at 7:45 am on Saturday, October 20th, her forefoot having passed inspection and the broken cable having been replaced. The Derwent made her run home in a quick-fire 88 days to Deal, eclipsing the run of the Achilles by six days. Despite both voyages being considered quite rapid, they were left in the shade by the voyage home of the Star of Italy. The crack clipper glided past Deal on Sunday, January 13th 1889, 85 days from Sydney. She even managed to beat the Cutty Sark's run home by one day. Kearney, Prior, McBarnett and Johnstone were home at last but not for long. They still had the rest of their indentured time left aboard the Star of Italy before they could consider themselves truly free.

The four young men completed their indentured time aboard the Star of Italy before each successfully sat for and gained their Second Mates certificates. Edward McBarnett then moved onto work as second mate upon the Star of Russia before making the move into steamers, finally gaining his master's ticket in 1899. Frank Kearney stayed on board the Star of Italy for a further nine months as third mate before transferring across to the Star of France as second mate in 1891 for another year. He then made the cross over into steamers to gain his mate's certificate there before again taking employment with Corry's as first officer of the Star of France for one more trip in

Star of Greece

1894. From there he was made second mate aboard the SS Star of New Zealand before qualifying for his master's ticket in 1897.

James Johnstone moved straight into steamers upon gaining his second mate's ticket going to work for the Glen Line of steam packets in the East Asia tea trade. His first berth was as fourth mate aboard the 3000-ton SS Glenfruin sailing from London to Japan and back. After three voyages he transferred to the SS Glenesk upon which he was second officer until 1897. He stayed with the Glen Liners until 1901 when he passed the ordinary ships master exams for square riggers. James then went onto a long and successful career in steamers distinguishing himself throughout the First World War qualifying for his steamer master's ticket in 1917.

Alfred Prior was another of the apprentices who stayed with Corry's once he'd completed his apprenticeship starting as bosun aboard the Star of Germany in 1890. He voyaged with her until September of 1893 finishing up his time with the ship as second mate. Upon gaining his first mate's ticket his career took a sudden shift as he moved into working for the Cuban SS Co Ltd, London, aboard their fore and aft rigged steamer the Cayo Romano. He started aboard her as third officer and kept on in the company's employ working aboard the SS Cayo Mono. Alf Prior then moved over to work for F.E. Bliss of London as third officer aboard the SS Potomac. With his first mate's certificate in hand, he was promoted to second officer aboard the Potomac and then the SS Glanton. After passing the Board of Trade square-rigger ordinary masters exams in 1899 Alf Prior stayed with steam building a successful career right through to the war.

The ordinary sailors that survived the Star of Greece were very much left to fend for themselves once they were discharged from Corry's service on July 21st 1888. All were exhausted and sported injuries about their heads and upper bodies from the savage struggle to the beach. The four men; James Revell, Carl Claeson, Edward Bluhm and David Bruce were all staying at the Prince Alfred Seaman's Home in Port Adelaide under the care of Emanuel Hounslow, head of the Mission to Seamen whilst the various government enquiries were taking place.

Star of Greece

The wooden barque 'Howard', 580 tons. Built 1864, Nova Scotia.
State Library of South Australia.

Star of Greece

None of the four could return to sea until they knew the outcome of the investigation, and each needed time to heal. The Star of Greece relief fund provided money for new clothes and seagoing outfits, sea chests and bedding for the sailors. In addition, each man received funds to cover living and other expenses whilst they waited for the local authorities to no longer require their presence ashore. James Revell was given £14 4s 6d, Carl Claeson £14 16s 2d, Edward Bluhm £13 16s 3d and David Bruce £17 3s 4d. The amounts handed out were roughly equivalent to two or three months sea pay and would have just covered their shore expenses even staying at the Seaman's Home. Their time ashore perhaps began to pall as money ran low and the goodwill of the locals began to wane. Henry Crowe eventually found the sailors berths on vessels homeward bound for England. Once in Britain Carl could head home to Gottenberg, Edward back to Hamburg, David to Montrose and James back to Ipswich, or not. Of the four he was the least enthusiastic about a voyage to Blighty.

Henry Crowe presented a rather imposing figure being employed by the Marine Board to procure sailors for outgoing vessels when they were short of crew. Born in 1829 he was a Crimean War veteran who had served at the battles of Sebastopol and Alma and proudly wore his medals and ribbons at official occasions. He had served as a merchant sailor for many years before arriving in Port Adelaide in 1857. He worked as a stevedore before being appointed to the position as official Seaman's Runner at the Seaman's Home in Port Adelaide in 1884. He was very good at his job, being very attentive to the needs of those men under his care. Unlike the average boarding house runner, Henry Crowe was held in high regard by the marine authorities at the port and could be relied upon to protect the interests of the Home and the sailors staying there. With Commerford, Bruce and Bluhm off to sea aboard the Guy Mannering, the ship that Crowe organised for the two remaining survivors was the 1063-ton Nova Scotian wooden barque Howard. She had just arrived from New York and under the command of Captain Alexander Crowe Vance. Captain Vance was a hard-driving Nova Scotian of the old school and upon the Howard's arrival most of her incoming crew had jumped ship and Captain Vance was in dire need of replacement sailors.

Star of Greece

The Howard dropped anchor on the afternoon of August 15th after a 113 day run from New York, loaded with general cargo and case oil, with Captain Vance's wife and young daughter also aboard. Upon discharge of her cargo at McLaren Wharf the clipper was shifted across to North Arm to begin loading wheat. The barque had been chartered by John Darling & Sons to carry a cargo of 13123 bags of wheat to London as a replacement cargo for that lost with the Star of Greece. Captain Vance approached the Seaman's Home in search of new crew for the run to London and Henry Crowe was happy to oblige. The two Star of Greece survivors were signed on in the second week of September, they were contracted at £6 a month plus a sign-on bonus. The Howard was cleared out on September 15th and towed out to the Semaphore anchorage to await her final clearances.

Captain Vance was all set to depart on Wednesday the 17th when he struck trouble with several members of the crew who refused to follow any of his commands. After several attempts to get the men to work, Vance had the chief mutineers Able Seamen William Kerr and James Revell placed in irons until police could be signalled to arrest the men and take them ashore. Once the two recalcitrant sailors were taken care of the rest of the foremast hands agreed to get back to work. As a result of this action, the Howard's departure was delayed by 48 hours. The trouble makers appeared in the Port Adelaide Magistrates Court at 1:00 pm the next day where Captain Vance charged them with disobeying his commands. Neither would give reasons for their conduct but happily plead guilty to the charge. As a result, they were sentenced to one month's gaol and fined two days' pay. With order at last restored the Howard set sail bound for London early the next day. The barque had a slow and tedious run home, arriving off of Deal on January 19th 1889, 152 days from Semaphore.

Of all the survivors from the Star of Greece, John Dashwood Hazeland had the most to lose. His testimony was key to determining how the ship was wrecked and who was to bear responsibility for the loss of a valuable asset and 18 of her crew. Upon leaving the apprentices to the tender mercies of Captain Cotter John Hazeland took time off to spend with family before boarding another Orient Line steamer for London.

261

Star of Greece

SS Rhipeus owned by China Shippers Company.
James Adamson, photographer, Rothesay.
University of Glasgow Archive Services.

Star of Greece

Despite being cleared of any culpability in regards to the loss of the Star of Greece by the Marine Board, and later lauded for his actions by the Select Committee Report Hazeland had left the colony with a cloud hanging over his character and conduct. Arriving back in London he had to face several hours of intense questioning by the Trinity House Masters who wanted know more about the hazards encountered and the way the Star of Greece was handled in the lead up to her foundering. The pilots of Trinity House were the lesser of John's problems as he headed firstly to Fenchurch Street, London and thence to Belfast to meet with Corry's management to explain how and why their flagship had been lost. His employers were satisfied with his answers and conduct for John Hazeland was placed as first mate aboard the Star of Bengal for another eight months before he passed the Board of Trade's Masters Exams at the end of 1889. John Hazeland then spent several more years working aboard various Corry steamers as second and first Officer whilst he learnt the arts and trade of being a steamship captain.

Despite his many years of service with the company, culminating in almost ten years as a first mate aboard Corry's various sail and steam vessels there remained a question mark over the man and John was never able to gain the much-desired promotion to ship's master. This all changed in 1902 when Captain Hazeland left Corry's and joined Alfred Holt and Company as an officer in their Blue Funnel Line of steamers. His first appointment was as master of the 1800-ton Blue Funnel steamer Priam bound from Liverpool to China. This was the start of a long and successful career with Holts that lasted for more than 30 years. Some of the Blue Funnel Liners commanded by John Hazeland included the steamers Rhipeus, Antenor, Cyclops, Telamon, Agamemnon, and the ss Ulysses. A Royal Navy Reserve Lieutenant he served in World War One as master of the armed merchant steamer Agamemnon. The ship was attacked with gunfire by the German submarine U 48, in 1917 off the south-west coast of Ireland, escaping at speed, whilst returning fire. John Howard Dashwood Hazeland swallowed the anchor in 1925 and lived on in retirement for many more years in Bristol, dying in 1941 at the age of 79.

Star of Greece

**The wedding of Isobel Hazeland and Geoffrey Lloyd in 1932.
Captain J.H.D. Hazeland back row, second from the left with wife,
bride, groom and bridal party.**
Western Daily Press 04 July 1

Star of Greece

End of an Era.

The loss of the Star of Greece proved to be a watershed moment for JP Corry's business model. No longer could they afford the ruinous premiums that many marine insurance companies charged to cover their blue water clippers. In just four years the company had suffered horrendous losses with the destruction of the Star of Scotia, the Star of Albion and now their flag carrier the Star of Greece, which had been uninsured. The age of sail was in its twilight and steamers were taking over more and more of the traditional routes once plied by fleets of clippers. The company had built its last windjammer in 1886, the 1781-ton steel-hulled Star of Austria specifically for the jute trade. This last tip-of-the-hat to the age of sail did not spend much time on the traditional East India route and was forced to tramp her way around the world seeking general cargoes.

JP Corry & Co soon began to divest themselves of their older, smaller sailing vessels. The first to go was the aging iron barque Star of Erin. Built-in 1862 the 1000 ton vessel was sold to Park Brothers of London, one of the few companies still willing to risk buying sailing ships and send them on the colonial runs. The little clipper lasted barely three years before coming to grief in New Zealand's Foveaux Straits. The Star of Erin left the harbour town of Bluff on the afternoon of February 6th 1892 loaded with oats, wool and tallow. Her master Captain Hopkins set course south by east attempting clear Waipapa Point on a freshening south-westerly breeze. By late evening conditions had worsened dramatically as a south-easterly gale roared in bringing with it low cloud and driving rain. Visibility dropped to less than one mile as the Star of Erin groped her way along the coast. At 11:30 pm the clipper ran hard aground gouging her way through Waipapa Reef. Waves immediately began sweeping the deck from stem to stern. Despite the dangers, Captain Hopkins knew it to be suicide to attempt to launch a boat and ordered the crew to remain aboard until daylight.

Star of Greece

Photograph of a painting of the Star of Erin, wrecked on the South Island of New Zealand, circa 1892.
Original artwork by J. Kerr. Larne Borough Council.

Star of Greece

In the early predawn light, the first boat got away with the mate and ten men aboard were rowed around to Fortrose to alert authorities and then returned to the stricken vessel with a local lighthouse keeper acting as a guide. Sea conditions worsened as the morning wore on and the boat returned to Fortrose taking with it the ship's papers and most of the crew's belongings. Captain Hopkins in a second boat stayed by the stricken vessel until 7:00 am when staying alongside proved too dangerous and the keeper of Waipapa Light House guided them safely through the reef to land. The crew of nineteen left just in time as by 8:00 am the masts had gone over the side and the Star of Erin had broken into three pieces much like the Star of Greece had back in '88. There were no deaths recorded in association with the wreck though the barque and her cargo were a total loss.

The subsequent enquiry laid the blame squarely at the feet of Captain Edward Lovett Hopkins, the former master of the Star of Erin. The authorities put the accident down to overconfidence on the captain's part in estimating the distance from Waipapa light resulting in the vessel running in too close to the shore and straight across Waipapa Reef. Unable to take sightings due to the thick weather Captain Hopkins had not been able to take a cross-bearing and was forced to rely upon dead reckoning to estimate the barques' position in the difficult conditions. The Marine Board felt that despite this there was little really that Edward Hopkins could have done differently and thus despite the stranding being his fault, the Board refused to remove his Master's Ticket or fine him. The Star of Erin was not the only former Corry clipper to come to a rather catastrophic end. However, hers whilst the first was also the least dangerous. Many other crew members coming to grief on former Irish Star vessels were not to be so lucky.

The Star of Denmark, sister ship to the Star of Erin was also sold in 1889 to F M Tuckers of London who retained the name of the little 1000 ton clipper. The new owners soon realised that she was not going to make them a great deal of money and so, on sold her to Hine Brothers' Holme Line of Workington. The Hine Brothers owned the Abbey Holme, SS Alne Holme, Brier Holme, SS Esk Holme, Eden Holme and the SS Glen Holme to name but a few. The Star of Denmark was renamed Denton Holme in 1890 and sailed from

Star of Greece

Glasgow on June 10th, with a load of iron pipes for the Water Board in Perth, Western Australia. The barque had almost reached her destination when on the morning of September 26th at 12:30 am she ran clean across a rocky shelf just to the north of Transit Reef. Her Master, Captain Rich had been attempting to sail the clipper around the north-eastern tip of Rottnest Island.

The Denton Holme was stuck fast upon the reef and weather conditions began to worsen prompting Captain John Hoare Rich to have the crew prepare to abandon ship if necessary. By 3:00 pm the steamers Cleopatra and Rescue had brought the crew of the Denton Holme ashore with all of their personal belongings. The weather came in thick and squally from the northwest that night as huge waves began to batter the stricken vessel. The Denton Holme eventually broke up into three pieces and little of her cargo was salvaged. The Marine Board enquiry found Captain Rich was at fault and suspended his Masters' ticket for three months. Captain Rich's luck would eventually run out when he along with all but one sailor, Oscar Larsen, were killed when the barque Brier Holme was wrecked on a reef north of Port Davey, Tasmania in 1904.

The third Corry vessel to be offloaded in 1889 was the aging Jane Porter, which was snapped up by William Ross and Co' for £5100 who then resold her to the Hamburg based firm of Hermann Burmester & Co', owners of the Oldenburg Line of vessels. The vessel was reduced to barque rig and renamed Nanny. Burmester's held onto the clipper until 1902 when they sold her to the Swedish shipping firm of Olsson Sten of Gothenburg who changed her name yet again to Trichera. The barque's gallant run lasted until the night of May 31st 1905 whilst anchored off the coast of Natal, South Africa. The Trichera, under command of Captain Hermansson, had sailed from Bunbury, Western Australia on April 10th loaded with hardwood sleepers for East London. Anchored in an open roadstead south of Aliwal Shoal opposite the village of Park Rynie the Trichera lay with both anchors set in the face of a hurricane roaring in from the east.

As the storm worsened the barometer dropped to 29.43 and winds picked up to 55 knots with much stronger gusts. Spume crested rollers drove in from the Indian Ocean rising and shortening as they crossed the outer bars of the Aliwal Shoal. The first mate aboard the

Star of Greece

Trichera decided to set out additional kedging anchors to keep the barque securely snugged down to ride out the blow. However, at 9:00 pm an enormous wave smashed against the bow of the vessel striking with enough force to tear the anchor cables clean out of the hawse holes parting them both at the windlass. It was the work of but a few minutes by the tidal forces of wind and waves for the Trichera to find herself hard aground beam on amongst the breakers.

There was little that the crew could do in the growing darkness except hang on for dear life until morning. The Trichera's 27-year-old master Captain Karl Johan Hermansson was of little use having been bedridden for some time due to illness. As dawn approached the crew took shelter within the flooded fo'c'sle and in the mizzen rigging. They noticed a crowd gathering upon the shore. The crew tried many times to get a rope ashore to no avail. There were nineteen men and boys aboard when the first of them attempted to make for the shore. Unable to get a boat over the side, seas breaking clean across the deck, the sailors were left with little choice but to strike for the beach.

In twos and threes, the sailors dropped from the bow and were quickly swept on to the beach through the roiling surf. There they were warmly greeted by local townsfolk who took them into care. Karl Hermansson and five members of his crew were not so lucky, being drowned either trying to get to shore or when the Trichera finally broke into three pieces as the waves pounded her into the sandy shore off Point Rynie. The end of the Trichera (ex-Jane Porter) was eerily similar to the loss of the Star of Greece. Whilst it was the three bulkhead design which had kept any of the Stars from burning due to coal fire it was the same design that doomed so many men and boys once the ships had run aground.

Sentiment counted for little at Corry's once the shift to steam had begun, and thus the company had few qualms in selling its other flag carrier the Star of Persia in 1893. The clipper was snapped up by the Hamburg based firm of C.M. Matzen & Co, who renamed the ship Edith and put her into the Zanzibar - South America - Portland, Oregon run. It was on a voyage in 1903 from Puget Sound to Port Pirie, South Australia carrying a cargo of timber in her holds and upon the deck, that disaster struck. At 3:00 pm on March 19th whilst sailing south by west during fine weather the Edith ran hard aground on a

Star of Greece

coral shoal named Nereus Reef. The limestone outcrop was located on the northern end of Lansdowne Bank between New Caledonia and the Chesterfield Islands, in the Coral Sea. The coral all but tore the bottom out of the ship and within ninety minutes the Edith had broken her back. The ship's master Captain Oertel ordered the crew to launch the lifeboats, each craft was provisioned with three weeks of food and water. The Captain and ten sailors managed to get away in one boat whilst the first mate Nicholas Kreuger and another seven sailors successfully escaped in another. The two boats stayed together throughout the first day and on into the night of the 19th but then became separated in the dark moonless seascape. Come morning the boat containing Captain Oertel, second mate Bauer, and crew members F.P. Anderson, Nordal, Charles Roland, Johansen, P. Raine, J. Cevley, C. Walters, W Rix, and O. Petersen the cabin boy, was nowhere to be seen.

First mate Kreuger sailed his boat towards the Solomon Islands. Fierce and contrary winds and high seas made life aboard the cutter very uncomfortable and prevented the little craft from making much headway. Those sailors in the mate's boat were H. Ludwig, F. Engelstrom, H. Carlson, W. Hartig, P. Weise, T. Christiansen, and G. Heise. They sailed slowly southwest for nine days before sighting another vessel. The desperate sailors pulled hard on the oars determined to reach the safety of the clipper before she sailed out of sight. The ship turned out to be the 1800 ton German full-rigger C. H. Wätjen out of Bremen.

If Kreuger and his men had thought they were safe they were in for a rude shock. Much to their dismay it turned out that the C. H. Wätjen had lost her rudder, masts, yards and sails during a cyclone several weeks before, bound from New York to Yokohama with a load of case oil. The dismantled wreck drifted about the Coral Sea for another 76 days under jury rig before help in the form of a coastal steamer came into view off Yule Island just three miles from the New Guinea coast. The steamer turned out to be the SS Moresby bound for Port Moresby 70 miles away.

Star of Greece

The crippled ship C. H. Wätjen is met by the steamer Moresby off Yule Island.
San Francisco Call, Volume 93, Number 181, 30 May 1903

Star of Greece

Barque Edith, formerly the Star of Persia. Wrecked 1903, New Caledonia.
Brodie Collection, State Library of Victoria.

Star of Greece

The crippled ship was towed to safety and the castaways transferred to the Moresby for the run back to Port Moresby. From New Guinea they were then transported by the ketch Pearl to Cooktown, arriving on May 29th 1903.

The boat containing Captain Oertel and his band of ten seamen sailed off into the night losing track of the first mate's cutter after the first night. Making excellent use of his sextant and boat compass the captain managed to guide his little craft successfully towards the Solomon Islands, reaching shore after twelve days of rough sailing. Making landfall at Montgomery Island the sailors found no water upon the deserted atoll. Setting forth once again they made good time to a small islet known as Good Duck Isle. Here they were met by local Solomon Islanders who were less than welcoming in their reaction to the arrival of the castaways. Despite their threatening actions the Islanders did not approach too closely allowing Captain Oertel and two men to get ashore and refill their water casks.

The sailors did not stay long and were soon again underway. Despite their replenished water supplies food was getting scarce and then sailors were forced to hunt for turtles and fish to supplement their meagre fair as they sailed onwards. They made landfall again at a place called Cape Pitt where they received a much warmer welcome from the local villagers. The people here were more familiar with Europeans and fed the exhausted sailors cooked yams, the first hot meal the grateful men had consumed in 16 days.

It was whilst the men were staying at the village that they encountered a local trader, Fred Ericson, from the small island of Gavutu. The grateful seamen spent their time either working with the local missionaries based on Gavutu or aboard the trading schooner owned by Ericson trading and transporting missionaries about the Solomons. After two months in this tropical paradise the supply steamer Titus arrived and Captain Oertel and eight sailors elected to return to civilisation. The survivor's stay at Gavutu had not been without incident; the ships carpenter/blacksmith Charles Roland had died of fever several weeks before the SS Titus arrived and another crewman decided to stay on and continue to work aboard Ericson's schooner.

Star of Greece

The Titus sailed away from the Solomons, making dozens of stops on her meandering voyage back to Sydney. Word of the survival of Captain Oertel and his men arrived ahead of the steamer when a cable was sent from Norfolk Island heralding their miraculous tale of survival. The Titus arrived in Sydney on June 28th 1903 bringing with her the remaining castaways from the Edith. The ship passed through Sydney Heads at 4:00 pm and was docked at Circular Quay at 7:00 pm that night. The men were met by the German Consul and representatives of the Mission to Seamen who together took the sailors into their care at the Sailors Home, Sydney. There they joined many of the sailors from the first mate's boat who had previously given them up for dead. With the last of the crew's safe arrival ashore the saga of the Edith/Star of Persia was finally at an end.

The Corry clipper Star of Germany was one of the last few to still fly the red heart flag at the end of the nineteenth century, but in the end, even she was no longer considered a viable commercial concern. After she arrived in Bristol in February of 1897 from Iquique, the ship was put up for auction. The vessel was quickly snapped up by Foley, Aikman & Co. of London for a mere £3300 who sensing the chance to turn a quick profit almost immediately put the Star of Germany on the market for £4000. They were to be sadly disappointed as the ship lay at anchor for the next seven months until she was sold for £2300 to William Keene Parrett, a silent partner of Foley, Aikman & Co, Parrett was a shipping broker and marine insurer, he immediately put the ship on the run to the west coast of South America. Her canny owner sent his latest acquisition to Cardiff to take on a load of coke and thence to Coquimbo for nitrates on her return run. The Star of Germany remained in the nitrate trade for the next few years until lady-luck deserted the clipper as one disaster after the next plagued her.

In February of 1900, sailing from Bangkok to Pisco, Peru, the ship, whilst anchored in the lee of Islas Ballestas dragged her anchors during a violent storm and drifted ashore. After discharging a large portion of her cargo, she was hauled off the sandbank by a tug. Stevedores and the crew finished unloading the rest of the cargo. The Star of Germany was towed up the coast to Callao to be dry-docked as she was seen to be taking water into her forepeak.

Star of Greece

Star of Germany moored at Port Adelaide.
State Library of South Australia.

Star of Greece

Marine surveyors found that the ship had sheered through 146 rivets in her bow section and it was going to be a major expense to have the cracked iron plates taken off, repaired, and rivets replaced.

Her troubles were far from over for it was on the run from London to Australia at the beginning of 1901 that disaster again found the Star of Germany as she rounded the Cape of Good Hope. Caught in a nightmarish storm the clipper was dismasted as a vicious squall slammed into her on March 19th. The main topmast carried away at the base of the doubling taking with it the main and mizzen topgallant masts. Also going by the board were the main upper and lower topsail yards, the main and mizzen topgallant yards and the royal yards. The main-cap was sprung and the mainsail yard was left hanging in the slings as the crane shattered under the weight of the fallen rigging. The crippled vessel was hove to until her crew could clear up the wreckage and she then limped her way under jury rig to Algoa Bay where the sorry looking clipper was docked for repairs at Port Elizabeth. With a lack of suitable spars available it was several months before the Star of Germany was again ready for sea and whilst under repair, she was sold for £6000 to W.A. Rainford & Co' of Liverpool, a subsidiary of Blundell & Rainford, shipping brokers and underwriters.

The Star of Germany with Captain Watt still at the helm was sent out to take on a load of coal from Newcastle N.S.W., for Taltal, Chile. The clipper eventually arrived in Newcastle on August 9th, with her mizzen topmast shorn off at the cap, 222 days from London. The following year again saw the ill-starred clipper limping along South Africa's east coast, sporting damage from a recent dusting in the Indian Ocean. Her long-time master Andrew Watt had relinquished command of the ship to become Lloyds agent in East London and was at the time travelling in her as a passenger. The ship dropped anchor in East London Bay, with her bulwarks missing, the two forward boats carried away, her decks swept clean and several sails in rags. The Star of Germany had to dock for extensive repairs and was several weeks at East London before getting underway for Newcastle, arriving in January 1903.

The year 1904 saw the Star of Germany again headed for Newcastle this time via Table Bay, Cape Town. From there she safely reached

Star of Greece

New South Wales to take on coal for South America. The now barque-rigged vessel departed Newcastle on November 24th, arriving at the Port of Mollendo, Peru in January 24th 1905, 64 days out. Her new owners, Rainford's, ran a tight budget for the ship attempting to keep overhead's down. The Star of Germany was now known as a hungry ship with minimal crew aboard. From Mollendo she set sail for Pisco to take on nitrates for the European markets. Then on her way north her crew mutinied and her master Captain Roberts was forced to put into Callao to discharge the ship's crew who refused to proceed in the ship, so dangerous did they consider her state of disrepair. The Star of Germany lay at anchor in Callao for many months as her owners attempted to offload the troublesome barque. It was not until February of 1906 that Norwegian shipping broker Alexander Bech of Tvedestrand took the rusting clipper off Rainsford's hands for £2500. A replacement captain and crew were sent out to bring the barque, now named 'Grid' into Mobile Alabama to discharge her nitrate cargo. She arrived in Mobile in early May and was placed into dry dock for a refit and clean.

Once ready for sea the Grid, resplendent with new name boards, a fresh paint job and sporting a Norwegian flag, set sail from Alabama with a deck cargo of pitch pine, bound for the shipyards of Buenos Aires. Sailing south by east the Grid ran headlong into a category 3 hurricane east of Bermuda. With the barometer dropping to 29.18 and winds blowing at more than 60 knots the crew of the Grid soon found themselves fighting for their lives. The storm careered slowly north its outer edge passing over the stricken barque which by now had again been dismasted and had sprung a serious leak from her recently repaired bow plates. The captain wisely decided to head for the relative safety of a nearby port and sailed back west towards the Caribbean.

The barque was a crippled messed as she limped slowly through stormy waters towards the Windward Isles. The open roadstead of Bridgetown, Barbados was reached on October 7th 1906 where the anchor was, at last, let go in the busy waters of Carlisle Bay. The Grid discharged her cargo in port and was later surveyed, with a damage report being relayed to Bech's back in Norway. Unwilling to part with any more money to repair their battered vessel, Alexander

Star of Greece

Bech abandoned the barque and she was condemned. The vessel was stripped and her fittings auctioned off before an enterprising local shipping company had the hull patched up. The former Star of Germany then spent the next thirty or so years as a rusting coal hulk moored off of Georgetown before being broken up for scrap just before the start of World War Two.

Of all of Corry's clippers, the one to gain the most notoriety in her post J P Corry life was the 1800 ton Star of Bengal. The ship along with the Star of France, Star of Russia and Star of Italy carried the Red Heart flag until 1898 when the last of the clippers was sold to foreign buyers. The ships upon discharging their cargos of coal at San Francisco in 1898 were sold, as-is, to J.J. Smith and immediately put onto the California to Hawaii run. Smiths at first registered the vessels under the Hawaiian flag before the country was annexed by the United States of America. Once registered as an American vessel the Star of Bengal traded regularly about the Pacific carrying coal and grain. This continued until 1906 when Smith's sold their clippers to the Alaska Packers Association. The company was the largest salmon fishing and canning operator between San Francisco and various fishing ports in the Alaska panhandle. The company also snapped up the Star of Italy, Star of Russia, and Star of France and it was this small collection of vessels that gave the Star Fleet its designation.

The Star of Bengal lasted but two years with the newly formed Star Fleet before coming to a rather messy and tragic end. At the start of the fishing season, the clipper loaded with canning equipment Chinese, Filipino and Japanese workers and a crew of hardy seamen would sail north to Alaska. There she would spend the season tied to a wharf awaiting her cargo of canned salmon. Then at the end of the salmon run the now heavily laden barque like all of her Star sisters would be towed out to sea through the maze of channels and islands from whence the fully loaded clipper would sail home to San Francisco for the winter.

The Star of Bengal left San Francisco on the 22nd of April 1908 bound for Fort Wrangell, Alaska. The 1908 season was a very successful one for the company and by early September the barque's three holds were filled with 52000 cases of salmon and hundreds of empty or

278

Star of Greece

near-empty oil and kerosene drums. Working aboard the Star of Bengal were three Filipino and twenty-eight Japanese packers who were soon joined by seventy-four Chinese cannery workers who were placed in the 'tween decks accommodation alongside ten European fishermen and a cook, for the cold and stormy voyage home. The barque was under the command of Captain Nicholas Wagner a man of vast experience and a veteran of the Alaska to Alameda run. He had under his watch eighteen sailors, two mates, the carpenter and a ships cook, most were of Scandinavian and Irish stock.

Unable to sail out of Wrangell Harbour on her own the Star of Bengal required a twelve to eighteen-hour tow through a maze of islands. The barque was tied by a 750' long hawser to a pair of tugs, the Kayak and Hattie Gage which would guide the heavily laden clipper out to sea. At 8:00 am the lead tug, Hattie Cage hauled the clipper clear of the wharf as Captain Wagner gave the order to set lower topsails and a headsail to help guide his ship out into the open channel. The Kayak took a position alongside the Star of Bengal helping to steer the vessel down the channel. The problem was that neither tug had been designed to tow heavy loads nor built for very rough weather. The three vessels proceeded slowly south by west under fine and sunny skies. Steaming through the Sumner Strait north around Prince of Wales Island the weather slowly worsened as the day wore on. Late September was the start of the storm season when gales would roar in from the freezing Barents Sea. The turn south around Point Bakea was made at 8:25 pm on the 19th. The trio sailed into a freshening south-easterly wind that began to raise a nasty chop in the narrow strait. Brisk rain squalls ranged up the channel bringing gusty winds in their wake.

Warren Island was passed on the port side at 10:00 pm as increasingly squally gales and a shortening swell indicated the approach of a nasty Pacific blow. The barometer dropped steadily as heavy mists rolled in from the south at times blocking one vessel from the view of those aboard the others. With the wind coming off the starboard bow the Star of Bengal was straining against the hawsers that kept her attached to the tugs. As winds strengthened the Kayak and Hattie Gage struggled to make headway, themselves being badly pummelled in the worsening conditions.

Star of Greece

The masters of both tugs were concerned by the fact that no land had been sighted for some time. They feared that all three were getting too close to the ragged cliffs and shoals of Coronation Island. The Star of Bengal with lower topsails, headsail and spanker set was heeled over on the port tack making leeway to the east as the three vessels came within the lee of Warren Island. There was a very genuine need for the tugs to steer their charge clear of Coronation Island and this meant both turning to bring the clipper on a more westerly course. The manoeuvre worked but the tiny steamers were struggling. In the inky darkness, neither tug skipper could see that the wind had come around and caught the Star of Bengal flat-aback. Panic gripped those aboard the barque as she began to make sternway towards Coronation Island's rocky shore.

Several crew members begged Captain Wagner to cast loose the Kayak's tow hawser to bring the vessel round onto the starboard tack so that the clipper could claw her way off a lee shore. Wagner, refused, stating that as master of the towed vessel he was not allowed to cut the tow cable. Neither tug could make any progress as lines of squalls barrelled in from the southwest. At the same time, the Star of Bengal began to drag both her escorts into ever-shallower waters. The sternward progress of the vessels saw them crossing into shoaling waters as the Kayaks boiler was shaken loose in the horrendous seas. Soon after her smokestack carried away and she was in danger of catching fire, foundering or having her boiler fires snuffed out. The Hattie Gage was in equal trouble, for the hawser tied to the Star of Bengal came under increasing strain. Two head stays were carried away by the forces playing upon the tow line and the weight of the barque threatened to pull the Hattie Gage right under. At 3:25 am, less than one hundred yards from the beach both tug captains ordered the tow hawsers to be cut away. Those aboard the tugs could see and hear combers breaking over a rocky shelf close inshore.

The sharp thud of breakers could be heard above the roaring wind by those aboard the Star of Bengal. The crew felt the tow line whip away as the barque lurched free of her tugs. Captain Wagner gave orders for both anchors to be let go. The Kayak and Hattie Gage steamed to Shipley Bay, 26 miles away.

Star of Greece

They were desperately seeking shelter whilst the barque hung on for several hours in the wild surf off Coronation Island. Those aboard the doomed vessel were astride a bucking bronco, her bow shifting rapidly up and down with the passing of each breaker. There was panic below decks and the captain was forced to order the crew to batten down the hatches keeping the cannery workers locked in the 'tween deck spaces. As dawn broke the crew of the Star of Bengal could make out no sign of the departed tugs, just continuous lines of combers rolling in from the sea. Wagner was furious at the two tug captains for leaving his vessel in such a perilous position, not knowing of their own badly damaged states. With the anchors failing to hold all on deck could see that there was not going to be a rescue any time soon.

With each successive wave, the clipper was edging closer to her demise. By 8:00 am it was clear that the crew would have to abandon ship. With more than one hundred people aboard their only chance of survival was if a line could be got ashore. Captain Wagner called all hands on deck and ordered the distribution of cork-filled life preservers to all passengers and crew. Most aboard could not swim and there was a great deal of panic building amongst the cannery workers. Many spoke little English and had trouble understanding the crew's instructions.

As the seamen were attempting to launch the first of six lifeboats a roller lifted the bow twelve feet into the air. As the boat was halfway over the side the iron hull of the clipper crashed down upon the little craft smashing it to flinders. Wagner called for volunteers to take another boat to the shore whilst the mate ordered oil to be poured through the hawse holes to calm the breakers as they passed the bow of the ship. The age's old device worked and the four volunteers, Olaf Hansen, Henry Lewald, Fred Matson and Frank Muir pulled hard upon the oars as if the devil was behind them. Almost to the beach the little boat was lifted by a breaking wave and dashed upon jagged rocks.

Star of Greece

Alaska Packers Association Cannery in Wrangell, Alaska, with Star of Bengal at the dock.
John N. Cobb Photograph Collection, Alaska State Library.

Star of Greece

The Star of Bengal anchored in Alaskan waters.
Alaska Packers Association Archives.

Star of Russia, circa 1899.
State Library of Victoria.

Star of Greece

Star of Bengal rigged as barque as part of Alaska Packers fleet.
Captain Harold D. Huycke, State Library of Queensland.

Star of Greece

The fragile craft came apart, dumping her four helpless occupants into the surf. Those aboard thought that the men had been lost, but one by one they emerged from the surf and dragged themselves to safety. A line was soon got ashore and fixed from the mainmast to a stout tree high up on the beach. The first man to attempt to use the breeches buoy, the ships carpenter, was flung from the seat by a giant comber and catapulted through the air to his death. Captain Wagner then ordered the cables slipped as it was obvious to him that the anchors were dragging and he hoped to get the Star of Bengal side on to the shore. Once the chains were released the clipper swung round, stern first. The dying screams of torn iron plates resounded as the Star of Bengal began to break up.

The death of the carpenter.
Seattle Post-Intelligencer.

With her stern firmly wedge upon the rocks and the guts torn out of her mid-section the barque soon began to break up. Iron beams from amid-ships were ripped apart like matchwood as the clipper ground herself to pieces. Most crew on the poop deck were pitched into the sea as the Star of Bengal struck and many made it to shore, the sea not yet filled with deadly wreckage. Most cannery workers were taking shelter below decks when the vessel struck and were trapped when the bow section was torn off. The water swirled in where the fo'c'sle and fore-mast had been. Many workers were swept out of the hold along with hundreds of empty oil drums and cases of salmon.

285

Star of Greece

The Star of Bengal in the surf against Coronation Island.
Seattle Post-Intelligencer.

Soon after the mid-section containing the central hold broke away and like the bow sank rapidly into eight fathoms of boiling green water. As the vessel broke up those who could swim had to wend their way through a hazardous obstacle course of wooden spars, oils drums, debris from the ship and 48-pound cases of canned salmon. Of the one hundred and forty-seven men and boys aboard, most were shredded by wreckage in the water or the rocks and reef as they tried to make the safety of the beach. In the end, only twenty-seven cannery workers and crew made it to shore alive. The remaining one hundred and eleven men and boys had died horribly as the ship was torn asunder.

The Star of Russia like the Star of Bengal was one of the last jute carriers to bear Corry's Red Heart flag. With the rush to get out of sail altogether they sold the clipper in March of 1898 to Shaw, Saville & Albion of London, a company who maintained a large fleet of steamers but also retained a few sailing ships for the routes where it was not economical to send a steamship. The Star of Russia was sold

Star of Greece

after she had suffered a severe dusting on a trip to San Francisco in 1897. The ship had been pushed onto her beam ends after being hit by a squall and series of rogue waves. The cargo had shifted and the crew worked frantically to thrown 60 tons of cargo overboard before the vessel was righted again. Also, her bulwarks were stove in, deck swept clean and two of her boats carried away. Soon after the Star of Russia was sold on to J.J. Moore of San Francisco in 1898 and sailed under the Hawaiian flag shortly before the island was annexed by the United States. The clipper was kept on the Pacific run carrying grain, sugar and coal between Hawaii and San Francisco.

The Star of Russia collided with the coastal brig Lurline off of Black Point, on October 14th 1898, after an uneventful journey from Antwerp to San Francisco. The clipper with a load of cement aboard brushed up hard against the Lurline as she swung at her anchor on a flood tide. Neither vessel suffered serious damage. Soon after the Star of Russia was snapped up along with her sisters by The Alaska Packers Association and put to good use for another 26 years sailing from Alameda to Alaska for the salmon season year after year. On the 29th of July 1905, the aging vessel ran aground on Chirikof Island, Alaska. She was soon floated off and towed to Alameda for repairs which came to the staggering cost of US$56,000. By 1926 the days of the salmon packing windjammers were done and the grand old lady was sold off to shipping brokers Burns, Philp & Co' acting as agents for the newly established steamship company Establissments Bailands. The company was set up to service French interests between Noumea and Indochina. The company also purchased the 1000 ton barque Star of Peru (ex-Himalaya) and both were destined to become store ships and coal hulks servicing the steamers on their runs to Noumea.

The clippers were loaded with timber and set sail from Tacoma for Apia, Samoa arriving just over a month later. Once their cargos were discharged the vessels were stripped of most of their gear and then towed across to New Caledonia. Upon arrival in Noumea, the Star of Russia was renamed La Perouse whilst the Star of Peru was reborn as the Bougainville. They spent the next three years anchored as cut down coal lighters with cranes where their masts had once been.

Star of Greece

The St. Joseph, steaming through Sydney Harbour bound for Noumea, with the hulk La Perouse, and the small inter-island motor ship Lolita, in tow.
The Sydney Morning Herald Wednesday 22 May 1929.

With the onset of the Great Depression, the two hulks were moved, the Bougainville being towed across the New Hebrides outer islands to spend the rest of her career as a copra storage hulk. The former full-rigged ship was still afloat and being used as a rusty and battered copra barge in 1948, eighty-five years after her keel was first laid down.

In early 1929 the La Perouse was hauled across the Coral Sea to Sydney loaded with copra in tow behind the French steamer St Joseph. Later she would be towed back to New Caledonia to resume her career as a floating copra bin in Noumea Harbour. The stripped-down clipper spent many a year serving as depot vessel, coal hulk and eventually as a copra hulk in the New Hebrides backwaters under the name Du Petit-Thouras. The aging hulk was eventually towed into Port Villa where she lay rusting away year after year. The elements finally got the better of her in 1953 when she quietly sank at her moorings where she rests more or less intact today.

The Star of Italy was another of Corry's clippers that ended up under Hawaiian registration when she was sold to J.J. Moore of San Francisco in 1898. Her greatest claim to fame was that she held the all-time record for the fastest run from London to Calcutta. Her run of 77 days beat that of the Cutty Sark by two days.

Star of Greece

Barque rigged Star of Italy, Tacoma, Washington.
Alaska Packers Association barque.
Alvin H. Waite Photographic Collection. University of Washington
Libraries.

The ship was put into the Colonial trade carrying coal, timber, grain and sugar about the Pacific until she was acquired by Pope & Talbot of San Francisco in 1903, and operated by the California Shipping Co. Looking to turn a quick profit the vessel was refitted and sold onto the Alaska Packers Association in 1904 for use as a cannery ship. Decommissioned in 1925 the Star of Italy had put in almost 20 years of faultless service with the company making the round trip from Alameda to Alaska every year without major mishap. She was laid up at San Francisco for two years before being sold to the Darling Island Stevedoring & Lighterage Co' in 1927 for use as a store ship and eventually was stripped down to become a coal-lighter supplying the Alaska Packer's steamships.

Star of Greece

In the 1930s she was re-rigged and loaded up to sail for the city of Buenaventura the main port of Colombia on the Pacific Coast. There the ship was placed under the Columbian flag having been purchased by a local company supplying coal from Columbia's interior to steamers that plied the coastal and Caribbean ports. The hulk upon arrival was again stripped of her upper workings with derricks being mounted upon her lower masts where spars and sails once had been. The former jute clipper, one of the fastest iron ships ever built ended her life as an anonymous rusting wreck whose fate was to be scrapped for her iron sometime during World War Two.

The last of the Harland & Wolff clippers laid down for Corry's was the Star of France, built in 1877. She was their most consistent performer and perhaps the best example of a Belfast built iron clipper to come out of the Queen's Island yards. 1898 saw the vessel bought up by J.J. Moore's Puget Sound Commercial Company for £6000 and put on the Hawaiian register. The Star of France stayed in the sugar and coal trade until 1900 when she came under the American flag and was put into the timber trade running softwoods from Port Townsend, Washington to various Australian ports before returning to Hawaii with Newcastle coal. Moore's kept the ship in her full rig for the next two years until she was picked up by the Pacific Colonial Company, and chartered to the Alaska Packers Association. The Packers put the ship through a complete refit and overhaul to accommodate 150 fishermen for the Alaskan salmon season.

The Star of France's career as a salmon-runner lasted until 1925 when she was superseded by steamers and the clipper was laid up at Alameda until 1928 when the ship was bought by Chicago millionaire John Borden who dreamt of outfitting the clipper for an around the world trip to be crewed by sea scouts. The scheme came to nothing and in 1933 the hulk was picked up by shipping broker Louis Rothenburg, of Oakland, California. He quickly on sold the ship to Norwegian-born Captain Joakim Andersen, the owner of the Hermosa Amusement Corporation. By the time the ship came to notice she was in a poor state of repair, her paint was flaking, hull rusting and rigging slack and strained.

Star of Greece

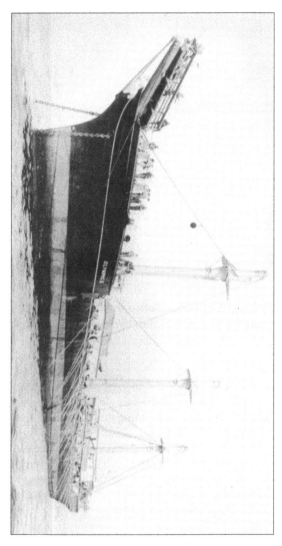

**Star of France now fishing barge Olympic II anchored off
of San Pedro, California, circa 1939.**
Los Angeles Public Library Archives.

Star of Greece

**The Nippon Yusen Kabushiki Kaisya owned steamer Sakito Maru
and the bowsprit of the Olympic II Circa 1940.**
Naval Historical Centre.

The company's owner, Captain Andersen named her the Olympic II and had the former clipper fitted out as a fishing barge, removing all her upper masts and yards, and cutting down her bulwarks. In early 1934, the Olympic II was towed down from San Francisco and moored off Hermosa Beach. For the next six years, the old lady was a popular attraction for tourists and anglers in South Bay. Overnight guests could be accommodated in seven staterooms when they weren't availing themselves of the concession stand, restaurant, nickel and two-bit slot machines or card games.

In 1940 Captain Andersen decided to move the Olympic II over to Horseshoe Kelp, a spot dangerously close to the shipping lanes, within sight of the lighthouse on the west breakwater of Los Angeles Harbour. On the morning of the September 4th there were 25 passengers and crew aboard the Olympic II looking forward to a

Star of Greece

lively day of fishing. Just after 7 a.m., the 9400-ton Sakito Maru, a freighter of the Japanese-owned Nippon Yusen Kabushiki Kaisya line emerged from the dense fog that covered the bay and ploughed into the port side of the Olympic II. Striking the former clipper amidships the momentum of the steamer drove her bow through 23 feet of the barge's 38-foot beam, smashing in the latter's side plates and snapping her keel.

The stationary vessel, in turn, was driven more than one hundred yards sideways. As the steamers' engines were placed all aback full the Sakito Maru reversed out of the breach. Thousands of gallons of seawater rushed into the Olympic II's hold, her water-tight transverse bulkheads having long since been removed. The lightly ballasted and now top-heavy barge was hard over on her beam ends and in less than two minutes rolled over and sank. The old hulk took with her to the muddy bottom eight people. Included amongst the casualties were the barge's keeper, Jack Greenwood, barge master and mechanic, concessionaire Joseph Karsh, Peter McGrath of Lynwood and his 9-year-old son, James, fisherman John Sylvester, and three local teenage boys, Curtiss Johnson, Peter Mayo and Joe Culp. Seventeen souls were saved by the coastguard and other nearby fishing and pleasure boats.

The loss of the Star of France marked the end of an era but the beginning of the end came about in much more mysterious and tragic circumstances. J P Corry's final clipper, the 1781 ton, steel-hulled Star of Austria was the last hurrah for the company before they made the switch to steam. Her first skipper was Captain Stewart Henry Willis formerly of the Star of Germany, who took her out to Calcutta where she loaded cotton, jute, rice, wheat, hemp and sacking for London. She visited Newcastle N.S.W in 1888 even as the Star of Greece was sailing towards her doom, to load coal for the west coast of South America. Captain Willis stayed with the clipper well into the 1890s and even though the Star of Austria had been slated for the East India trade she rarely went there. Under Willis, she visited Cardiff, Shields, Adelaide, Melbourne, Rangoon, San Francisco, Hamburg, Valparaiso, Lyttelton, Astoria and thence back to London and Queenstown. The ship stayed under Captain Willis' care until 1893 when he resigned to take up a position as a marine surveyor for

Star of Greece

the Canterbury Marine Underwriters Association. Corry's then passed command over to Captain Samuel James Russell.

Sam Russell was a long time Corry company man having served as both apprentice and officer aboard a variety of their ships before gaining command of the Star of Persia. He had the added advantage of being the son-in-law of Captain William Shaw, Corry's fleet Commodore, master of the 3000 ton Star of Victoria and first captain of the Star of Greece. Captain Russell had aboard with him his bride Mary who proceeded to make the saloon accommodation into a home. The officers and crew appreciated her gentle touch which was in sharp contrast to the hard-driving, no-nonsense work ethic and attitude of her husband.

From the start, the Star of Austria under Russell's command was put onto a new route to collect copper ore from the Baja Peninsula of northern Mexico. The ship left Newcastle U.K. on June 9th 1893 for Santa Rosalia, Mexico, passing St Catherine's Point on the 12th and after a record run reached the dusty outport, 103 days from the Tynemouth. Such was the adversity of weather off the Horn that Captain Russell was forced to travel 150 miles south almost to the ice barrier, where the crew suffered horribly from the cold, to make westing around the Cape. The Star of Austria set a record which had yet to be broken by a clipper travelling from the Tyne to Santa Rosalia.

The port of Santa Rosalia was a small copper mining settlement situated on a dusty and barren shore overlooking the northern reaches of the Sea of Cortez. The mine was opened up in 1868 but was considered a minor and unprofitable venture until a French mining firm, Compagnie du Boleo moved into the area in 1885. They established an open cut pit mine and built the town and an ore processing plant around it right upon the shore. The Mexican government at the time hoped that the mine would bring development and money into the arid and unpopulated region.

Star of Greece

Santa Rosalia, Baja California, Mexico c.1900.
U.S. Naval Historical Centre.

Star of Greece

The mine proved highly profitable and soon Chinese, Japanese, Yaqui Indians and Mestizos were brought in to work in the mine. Many died of malaria, typhoid and yellow fever or accidents due to unsafe and unhygienic working and living conditions. The company created an artificial harbour from the slag of the smelter which allowed ships to tie up to the rocky pier for handloading by local stevedores. Soon ships from all over the world were visiting the port to load up with the extremely high-grade ore.

Corry's soon picked up a charter to take coal for the smelter at Santa Rosalia and thence copper ore to Astoria. It was on this run that Captain Russell and his clipper set yet another speed record. Once her holds were full the Star of Austria cracked on down the Sea of Cortez and around the bend to Astoria, Oregon to take on a load oo wheat. Her run of 29 days set a record that has yet to be broken, proving beyond doubt the sailing capabilities of the clipper and her canny master.

The Star of Austria arrived in Astoria on December 20th 1893, just in time to avoid losing her precious grain charter. The stevedores of Astoria were highly efficient and the ship completed her loading by January 12th 1894. Captain Russell had the charter to take 2000 tons of wheat to Hull departing on the 14th. After a slow voyage punctuated by light and fickle winds, she picked up her pilot off of the Isle of Wight on May 28th, 134 days from Oregon. It was whilst they were on the return voyage that Mary Russell announced to Sam that she was expecting their second child. The couple already had one son, William who had been born in Belfast in 1891 yet was growing up aboard the Star of Austria. He was a favourite of the crew who let him have the run of the aft deck in fine weather. Upon discharge of her valuable cargo, the Star of Austria was put on the slip for a shave and a paint with antifouling agents before being refit for a tow around to Cardiff for yet another load of coal for Santa Rosalia's smelter.

Star of Greece

Mozambique, 2244 tons. Built at Pt. Glasgow, 1892.
Brodie Collection, La Trobe Picture Collection, State Library of
Victoria.

Star of Greece

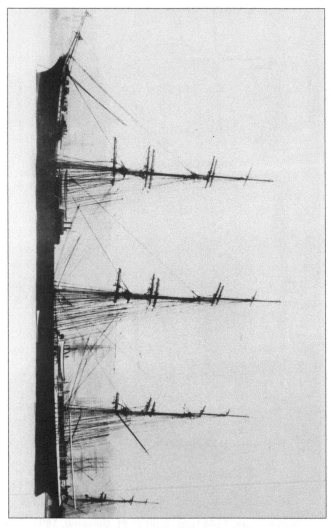

**The Star of Austria, Built 1886 by Workman, Clark and Co., Belfast.
Lost off Cape Horn, circa 1895.**
State Library of South Australia.

Star of Greece

The Star of Austria departed Cardiff on July 20th 1894 and left her tug off Lundy Island, at the mouth of the Bristol Channel later that evening. She crossed the equator on August 15th just 26 days out. This time she had a slow and tedious trip around Cape Horn having to avoid many giant icebergs that were still there from the previous season. The clipper experienced light and contrary winds for much of her run up the through the Pacific and did not arrive at Santa Rosalia until New Year's Eve, after 164 days at sea. The Star of Austria and her crew were in for a long wait as labour shortages and increased demand for the mine's ore meant a long, hot lay-up in the crowded roadstead. The mine had recently been sold to an American syndicate for $100000 and was undergoing a period of reorganisation. The Star of Austria was soon joined by the four-masted barque Mozambique which had left Cardiff on August 29th, long after the Corry ship. She arrived on January 16th, 140 days out from Lundy. Those aboard were forced to wait for almost three months for the hold to be filled.

The Star of Austria was cleared out from Santa Rosalia on March 23rd 1895 having gained a charter to head to Queenstown for orders. She had aboard almost 2000 tons of copper ingots and bagged copper ore causing the heavily laden ship to ride well down on her marks. The vessel was well manned with a crew of twenty-nine aboard, including Captain Russell, three certificated officers, a boatswain, carpenter, sailmaker, steward, cook, six apprentices and fourteen Able Seamen in the fo'c'sle. The Star of Austria received her final clearances on the morning of March 25th with Russell giving the order to cast off from the mooring buoy that same morning. The winds up to the time of departure had been from the southeast bringing with them lines of dust-laden squalls and thunderstorms.

By the day of departure, the wind had swung around to the northwest blowing away the inclement weather. The Star of Austria set sail down the Sea of Cortez as the wind veered to the northeast bringing low cloud and haze. The barometer was at 29.90 and falling slowly indicating a gradual change was approaching. In the narrow and reef-strewn waters of the narrow sea, Captain Russell kept his ship's upper and lower topsails set with headsails and spanker in expectation of a freshening south-westerly gale. The storms and rain swept in from the

Star of Greece

Pacific on the 28th as the Star of Austria hummed along, east of Isla San Jose. Russell gave the order to shorten sail lest his ship be caught in irons and driven upon a lee shore.

The cold front brought with it rising seas and cold, hail filled squalls which battered the clipper as she sailed down the gulf. Winds continued to freshen from the south as the vessel attempted to weather the approaches to Cabo San Lucas on March 31st. It was slow going as the ship tacked her way through the mouth of the Gulf of California and out into the open sea, passing the four-masted barque Mozambique, on April 1st as she was on her way to Astoria to load wheat. This was the last confirmed sighting of the Star of Austria as Captain Russell piled on all sail and cracked on down through the Pacific on his way to Cape Horn.

The Star of Austria encountered fair weather for much of the voyage south. Most voyages that year around Cape Horn were long and tedious in either direction, and many vessels were unreasonably delayed. Insurance premiums on dozens of ships rose sharply as more and more were reported overdue in both London and San Francisco. The average sailing time to the Horn from California was 60 to 70 days and many vessels were beset by a series of horrendous Cape Horn Snorters with their hurricane-like winds and mountainous greybeards. The British barque Talus rounding the Cape from the east at about the same time as the Star of Austria was passing to the west, encountered howling winds, and thunderous snow-filled squalls. At one point well south of Cape Horn the Talus encountered a series of giant waves that swamped the barque, sweeping her decks clean, stoving in her bulwarks and carrying away a ship's boat.

Another vessel, the Arabia struck the same series of storms on May 25th. The ship battled huge seas and gales before being almost completely dismasted. The battered vessel rolled violently her deck often completely awash as seams opened and the Arabia began to founder. The crew took to the boats and after a week being tossed about in the tempest managed to make the relative safety of Diego Ramirez Island. The Star of Austria was also caught in the same stormy conditions yet those aboard her were nowhere near as fortunate as the crews of the Talus and Arabia. At some point during

Star of Greece

the worst of the Cape Horn hurricane, the Star of Austria foundered and sank taking all aboard to the bottom.

Those lost were Captain Samuel Russell 33, Mary Russell 26, William Russell 4, James Russell 1, First Mate John Saul 45, Second Mate William Thompson 24, Third Mate Charles Armstrong 26, Boatswain John Forbes 42, Sailmaker David Lawson 42, Carpenter William Cumming 37, Steward Fred Stephens 20, Cook Charles Erskine 30, Apprentices; Arthur Peacock 19, Henry Bowden 19, Walter Holland 16, Bertram Newton 17, James Wilson 14, Thomas Cunningham 15, Able Seamen; James McDonald 28, Peter Power 23, Lars Larson 35, John Marlow 26, Felix Gall 30, Charles Hamilton 41, Jon Kelsted 29, David James 20, William Humphreys 22, M. Keating 25, George McAteer 25, B.W. Leighton 48, William Barnett 32. There was one sailor, Grust Sodehohn, who could count his lucky stars having signed off from the Star of Austria before she sailed.

The loss of the Star of Austria did not become readily apparent for several months after her departure. Many vessels coming south from around Cape Horn were making longer than expected passages due to the malignant weather in that region of the world in the winter of 1895. The previous two years had seen many a vessel lost in the region due to icebergs however this winter the storms were much worse than normal. Another Belfast vessel, Lord Downton, a four-masted barque of 2,262 tons, owned by Thomas Dixon & Sons, timber merchants of Belfast, was also missing as was the 2000 ton four-masted barque Stoneleigh, the French steel barque Marie Alice, the four-masted barque Lord Spencer and the Shaw, Savill, and Albion Company's freezing ship Timaru, which had left Melbourne for London on the 4th of June 1895 with a large quantity of wool and meat.

It was not until November that Lloyds of London posted the Star of Austria as missing. Corry's still held out some small chance that their ship and her crew would turn up but as Christmas approached all hope began to fade. Reports came in from arriving crews that wreckage matching that of the missing clipper had been found in iceberg filled waters off Cape Horn. With this heart-wrenching news, there was little doubt remaining as to the fate of the Star of Austria and those

Star of Greece

aboard. The passengers and crew of the missing vessel were officially declared dead at the start of December 1895 and a subscription was got up in Belfast to raise money for the less well-off families of the lost. In the end over £500 was raised and distributed to the parents, wives and children of the missing men and boys. Nothing could assuage the anguish of those left behind.

One man who felt the loss of the Star of Austria more than most was Captain William Shaw, master of the Corry steamer Star of Victoria. The ship was on a voyage from London the Melbourne, Sydney and thence to New Zealand and it was not until the vessel arrived back in London in November that Captain Shaw, and his son William, who was second mate of the steamer, learnt of the deaths of Mary, her husband Samuel and his two grandchildren William and James. The tragedy hit him hard, for in that time, aged 61, he swallowed the anchor and retired from life at sea. Captain William James Shaw, the first master of the Star of Greece, stricken with grief, moved back to the family home in Belfast. He lived quietly with his eldest daughter Margaret and her family until he died after a short illness in 1899 aged 65. His passing marked the end of an era for JP Corry's and brought to a close the story of the Star of Greece and her owner's involvement in the age of sail.

Star of Greece

Some sailed over the ocean on ships,

Earning their living on the seas.

They saw what the Lord can do,

His wonderful acts on the sea.

He calms the storm,

And the waves become quiet,

And he brought them safe to the Port they wanted.

They must thank the Lord for his constant Love.

PSALM 10

Star of Greece

Crew of the Star of Greece 13th July 1888

Name	Age	Position	Place of Birth	Last Residence	Fate
Henry Russel Harrower†	29	Ships Master	Broughty Ferry, Dundee Forfar	Picton Villas, Dartmouth, Parkhill London	Deceased
John Howard Dashwood Hazeland	25	First Mate R.N.R.	Balmain, Sydney	Sunnycroft, Pinhoe, Exeter	Survived
William Robert Waugh†	23	Second Mate	County Down, Donaghadee	Woodstock Road, Belfast	Deceased
Charles Martin Commerford	21	Third Mate	Ramsgate, Kent	Winstanley Villas, Ramsgate, Kent	Survived
William Parker†	41	Boatswain	London	High Street, Poplar	Deceased
Robert Donald†	27	Carpenter	Carrickfergus	Green St Carrickfergus	Deceased
George Coldrey Blackman†	36	Cook & Steward	London	Park Villas, Chadwell Heath Romford. Essex	Deceased
George Percy Carder†	19	Cabin Boy & Assistant Cook	Romford, Essex	Western Rd Romford. Essex	Deceased
Gustaf Carlson†	52	Sail Maker	Stockholm	Commercial Street London	Deceased
Francis 'Frank' Kearney	19	Apprentice	Queenstown, County Cork	Monkstown, County Cork	Survived
Edward James McBarnett	19	Apprentice	Wellington, New Zealand	Harrington Road, London	Survived
Alfred Prior	17	Apprentice	Brancaster, Norfolk	Townsend, Dover	Survived
James Johnstone	17	Apprentice	Lockerbie, Dumfries	Torwood, Lockerbie	Survived
John Airzee†	28	Able Seaman	London	London	Deceased

Star of Greece

John Gatis†	50	Able Seaman	Shildon County Durham	Ferry Sreet, Montrose	Deceased
Andrew Blair†	19	Able Seaman	Blythe	Montrose	Deceased
Peter De Smet†	39	Able Seaman	Ostend, Belgium	Ostend, Belgium	Deceased
Carl Claeson	21	Able Seaman	Gottenberg, Sweden.	Gottenberg, Sweden.	Survived
Alfred Owen Organ†	40	Able Seaman	St Martins, Middlesex	London	Deceased
George Irvine†	43	Able Seaman	South Voe, Dunrossness, Shetland	Dunrossness, Shetland	Deceased
Wilhelm Oerschmid †	27	Able Seaman	Hamburg, Germany	Hamburg, Germany	Deceased
David Bruce	24	Able Seaman	Montrose, Angus	High Street, Montrose	Survived
William James Miles†	31	Able Seaman	Middlesex	Blechynden Street, Kensington, London	Deceased
James Thomas Cattermole Revell	43	Able Seaman	Ipswich, Suffolk	Ipswich, Suffolk	Survived
Henry James Richard Corke†	17	Able Seaman	Portsmouth, Hampshire	Somers Road, Portsea Island, Hampshire	Deceased
Eduard Bluhm	24	Able Seaman	Libau, Latvia	Kronenstrasse, Libau, Latvia	Survived
Robert Muir†	19	Cooks Assistant (working passage home)	Glasgow	Glasgow	Deceased
John McVicars aka (John Rowlyn) †	40	Deckhand (working passage home)	Baths, Edinburgh	Port Adelaide. Formerly Couper Street, North Leith	Deceased

Star of Greece

1. Marine survey of the wreck site.
2. Headstone of Alice & James Bishop.
3. Painting of The Star of Greece sailing by moonlight.
4. The 2018 bronze plaque remembering the forgotten.
5. Paintings and photographs of the Star of Greece.

Star of Greece

Marine survey of the wreck site.

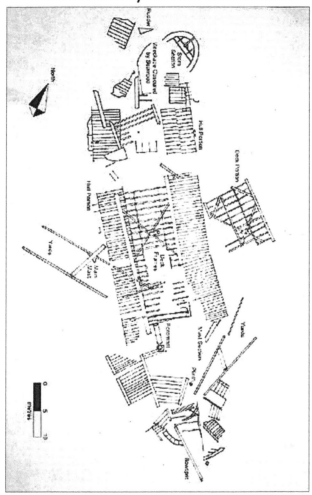

Star of Greece Wreck Site Plan 2003
A.Ash, J. Cooper and D. Cowan

Star of Greece

Headstone of Alice & James Bishop.

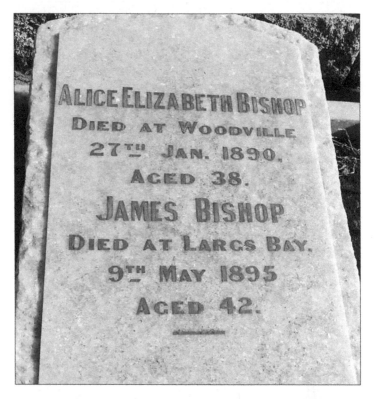

ALICE ELIZABETH BISHOP
DIED AT WOODVILLE
27TH JAN. 1890,
AGED 38.
JAMES BISHOP
DIED AT LARGS BAY.
9TH MAY 1895
AGED 42.

The joint grave of Alice and James Bishop buried together at
Woodville Cemetery, Adelaide, South Australia.
Alice died of burns having fallen into an open fire whilst cooking at
home in Woodville. James died of TB just 5 years later leaving the
Bishop children to be raised by family friends.

Star of Greece

The Star of Greece sailing by moonlight.

Star of Greece

The 2018 bronze plaque remembering the forgotten.

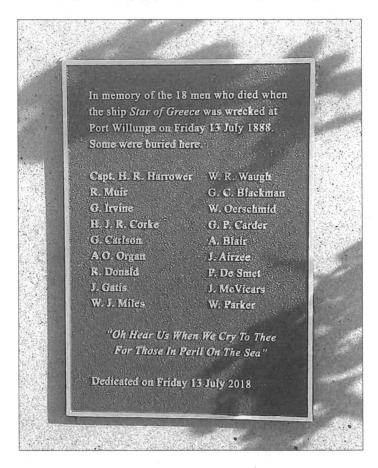

In memory of the 18 men who died when
the ship *Star of Greece* was wrecked at
Port Willunga on Friday 13 July 1888.
Some were buried here.

Capt. H. R. Harrower	W. R. Waugh
R. Muir	G. C. Blackman
G. Irvine	W. Oerschmid
H. J. R. Corke	G. P. Carder
G. Carlson	A. Blair
A.O. Organ	J. Airzee
R. Donald	P. De Smet
J. Gatis	J. McVicars
W. J. Miles	W. Parker

*"Oh Hear Us When We Cry To Thee
For Those In Peril On The Sea"*

Dedicated on Friday 13 July 2018

On July 13th 2018, thanks to the unstinting work of the
Willunga Branch of the National Trust the names of all who
were known to have died aboard the Star of Greece were
unveiled on a plaque placed next to the original memorial.
Hopefully, at last, the lost and forgotten can rest knowing their
names are now restored to history.

Star of Greece

Painting of the Star of Greece.

Artist Unknown.
State Library of Victoria.

Star of Greece

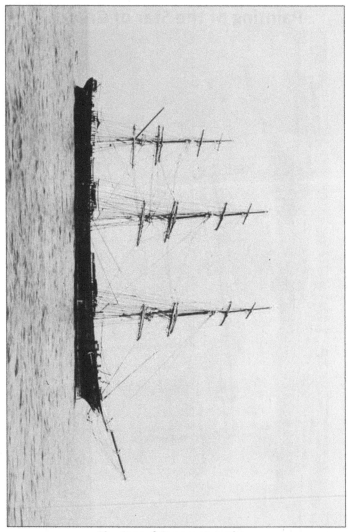

The Star of Greece at anchor.
The Advertiser Newspaper, circa 1938.

Star of Greece

The stern section of the Star of Greece.

Star of Greece

Bow section of the Star of Greece.

Star of Greece

Bibliography

Books & Publications

(1860 – 1998). **Lloyds Shipping Register**, Lloyds of London, United Kingdom.

(1888), South Australian Parliamentary Papers, August 14th 1888, III, paper 58 **Report of the Parliamentary Select Committee on the Wreck of the Star of Greece** with minutes etc., House of Assembly, South Australia.

(2018), **Star of Greece and other shipwrecks of the tragic shore**, Willunga Branch of the National Trust of South Australia, Printed by Office & Image, Willunga.

Ash, Aidan (2007), **The Maritime Cultural Landscape of Port Willunga**, South Australia. Department of Archaeology, Flinders University, Adelaide.

Blechynden, Kathleen, (1905). **Calcutta: Past and Present**. Thacker, Spink & Co, Calcutta.

Brereton, Robert Maitland, (1908). **Reminiscences of an Old English Civil Engineer 1858 – 1908**. Irwin-Hodson Company, Portland, Oregon.

Carter, H.C., (1938). **Jute and its Manufacture. John Bale**. Sons & Danielson Ltd. Oxford House, London.

Chaudhury, Nibaran Chandra (1921). **Jute in Bengal**. W. Newman & Co, Calcutta.

Dalton, Anthony (2010). **A Long, Dangerous Coastline: Shipwreck Tales from Alaska to California**. Heritage House Publishing Co', Toronto.

Eastwick, Edward B. (1882). **Handbook of the Bengal Presidency with an Account of Calcutta City 1814 - 1883**, Bradbury, Agnew & Co, London.

Chevers, Norman M.D., (1864), **Health of Seamen**, Military Orphan Press, Calcutta.

Christopher, P. (1990). **South Australian Shipwrecks: A Database 1802 – 1989**. Society for Underwater Historical Research, Adelaide.

Gennadios, Ioannes (1902), **Stephen A. Ralli – A biographical memoir**, Privatley Printed, London.

315

Star of Greece

Leng, Sir John (1895). **Letters from India and Ceylon**. John Leng & Co. Dundee.

Lloyd, S.K. (2001). **Wreck of the Star of Bengal**. The Sea Chest, Vol 35, Number 2 December 2001, Journal of the Puget Sound Maritime Historical Society, Seattle.

Lubbock, Alfred Basil (1902), **Round the Horn before the Mast**. John Murray, London.

Lubbock, Alfred Basil (1914), **The China Clippers**. Brown, Son and Ferguson, Glasgow.

Lubbock, Alfred Basil (1921), **The Colonial Clippers**. Brown, Son and Ferguson, Glasgow.

Lubbock, Alfred Basil (1922), **The Blackwall Frigates**. James Brown and Son, Glasgow.

Lubbock, Alfred Basil (1927), **The Last of the Windjammers, Vol 1**, James Brown and Son, Glasgow.

Lubbock, Alfred Basil (1932), **The Nitrate Clippers**. Brown, Son and Ferguson, Glasgow.

Lubbock, Alfred Basil (1933), **The Best of Sail, The romance of the clipper ships; an abridged edition of 'Sail', volumes I, II and III**. Pictured by the late J. Spurling, storied by Basil Lubbock, edited by Alexander Campbell, New York, The Macmillan Company.

Lubbock, Alfred Basil (1935). **The Last of the Windjammers, Vol 2**. James Brown and Son, Glasgow.

MacArthur, Walter (1929). **Last Days of Sail on the West Coast**, James H Barry Company, San Francisco.

Manning, G. (1988). **The Tragic Shore**. National Trust of South Australia, Willunga Branch.

Massey, Montague (1918). **Recollections of Calcutta for over half a Century**, Thacker, Spink & Co, Calcutta.

Murray, John, (1882). Milowka, Agnes, Kalinowski, David. VanZandt, David M. (2007), **Site Inspection Report Star of Greece**, Flinders University, Adelaide, South Australia.

Rothery, H.C. (1883). **Report into the Wreck of the Dunstaffnage**, Board of Trade, Liverpool.

Runciman, Walter (1905), **Windjammers and Sea Tramps**. Walter Scott Publishing Co' Ltd. New York.

Sexton, R. (1982). **Before the Wind: The Saga of the Star of Greece**. Australasian Maritime Historical Society, Gillingham Printers.

Star of Greece

Vernon-Harcourt, L.F. (1896). **Report on the River Hugli.** Caledonian Steam Printing Works, Calcutta.

Woodhouse, T., Kilgour, P. (1921), **The Jute Industry from Seed to Finished Cloth.** Sir Isaac Pitman & Sons Ltd. Parker Street, Kingsway.

Newspapers

Aberdeen Journal: 13 September 1876. 08 November 1877. 30 October, 20 December 1878. 03 July, 11 September, 12 September, 07 October 1879. 14 October 1881. 13 September 1876. 25 April, 27 September, 07 November, 08 November, 09 November 1877. 29 August, 20 December, 21 December 1878. 03 July, 11 September, 12 September, 13 September, 07 October, 06 November, 11 December, 13 December, 22 December 1879. 16 January, 07 October, 08 October, 23 December 1880. September 28, November 07 1895.

Aberdeen Evening Express: 12 September, 03 October 1879

Arbroath Herald and Advertiser for the Montrose Burghs: April 25, June 13, 1895

Adelaide Observer: 4 December 1880. 14 May 1881. 29 March 1884, 14 July, 21 July, 1888. 6 June, 13 June, 4 July 1903. 29 October 1904

Advocate (Burnie, Tas.: 1890 - 1954): 22 June 1926

Armidale Express and New England General Advertiser: 18 May 1877

Auckland Star: 7 May 1873

Belfast Morning News: 24 June, 01 July, 21 September, 21 December, 1868. 01 February 1869. 23 January, 14 June, 22 November 1871

Belfast News Letter: 24 June 1868. 03 March, 07 April, 21 June, 26 July, 27 July, 1869. 19 March, 22 June, 30 September, 18 November, 21 November 1870. 16 January, 20 January, 21 January, 16 December 1871. 02 January, 30 April, 23 May, 23 May, 30 August, 02 October, 09 December, 14 December, 16 December 1872. 04 January, 13 January, 15 January, 23 April, 19 May, 27 November 1873. 10 January, 04 February, 05 February, 12 March, 23 May, 25 May, 01 June, 08 June, 04 July, 06 October, 27 October, 04 November, 26 November 1874. 19 January, 20 January, 06 March, 09 March, 23 June, 05 November , 06 November, 08 November 1875. 26 January, 12 February, 09 March, 04 April, 05 April, 24 June, 27 June, 28 June, 29 June, 22 November 1876. 03 January, 16 February, 20 February, 23 February, 01 March, 24 March, 07 April, 17 April, 24 April, 04 September, 25 October, 07 November, 09 November, 10 November, 15 December 1877. 01 March 1878. 10 April, 11 April, 28 May, 10 September, 02 October 1879. 20 January, 01 February, 27 August, 30 August, 31 August, 03 September, 12 September, 15 October 20 October, 30 December, 1881. 20 January, 13 April, 25 May, 13 June, 24 November 1882. 16 May, 19

Star of Greece

October, 03 November 1883. 06 February, 05 March, 06 March, 24 March, 26 March, 27 March, 03 May, 23 June 1884. 08 July 1885. 09 February, 02 March, 22 March, 24 March, 22 April, 28 April, 03 May, 06 May, 26 May, 02 June, 03 July, 04 August, 06 November, 06 December, 24 December 1886. 01 July, 30 August, 30 November, 19 December 1887. 13 February, 20 March, 04 April, 12 June, 20 July, 23 July, 07 August, 18 August, 29 August, 12 September 1888. 16 July, 26 September 1889. October 23, December 25 1893. 22 November, December 13, December 25 1895. 12 March, 05 July, 14 October 1897.

Bendigo Advertiser: 15 May 1889

Biggleswade Chronicle: 14 December 1900

Brisbane Courier: 24 June 1875. 25 January 1884. 24 July 1928.

Border Watch (Mount Gambier, SA: 1861 - 1954): 18 July 1888. 1 July 1903.

Buckingham Advertiser and Free Press: 10 March 1877

Bury Free Press: 10 March 1877

Clarence and Richmond Examiner and New England Advertiser: 24 July 1888

Clerkenwell News: 17 November 1870. 21 August 1871

Cornishman: 07 October, 14 October 1880, 31 March, 23 June 1881. 26 January, 27 April, 25 May, 20 July, 07 December 1882. 08 November, 15 November 1883. 12 June, 07 August, 25 December 1884. 01 January, 26 February, 31 December 1885. 25 March, 13 May, 01 July, 02 September, 25 November 1886. 24 February, 03 March, 19 May, 01 September 1887. July 26, 13 September 1894. 03 March, 24 March 1898. 05 April 1900. 28 March 1901.

Daily Commercial News and Shipping List: 23 September 1893, 4 February 1899. 7 June 1905. 27 August 1907. 2 June, 9 June 1926. 16 January 1934. 19 October 1943.

Daily Alta California: 7 March, 22 March, 8 April, 5 May, 16 May, 23 September, 24 September, 25 September, 23 October, 11 November, 14 November, 15 November 1884. 29 March, 31 March, 3 April 1885.

Daily News (Perth, WA: 1882 - 1950): 25 September, 26 September, 2 October, 6 October 1890. 4 February 18993 June 1926.

Daily Telegraph & Courier: 20 November 1871

Dundee Courier: 27 July 1869. 23 May, 16 December 1871. 31 July 1872. 15 August, 28 August 1873. 12 March 1874. 28 May, 24 September 1875. 17 January, 18 January, 31 January, 01 February, 14 February, 24 June, 27 June, 05 September, 23 November, 24 November 1876. 15 February, 16 February, 01 March, 02 March, 07 March, 04 July, 06 July, 03 September 1877. 27 February, 30 May, 10 July, 12 July, 07 August , August 09, 12 August, August 13, 29 October, 12 November, 13 November, 15 November, 06 December, 09 December 1878. 05 March, 10 April, 12 September 1879. 22

318

Star of Greece

April, 06 October 1880. 01 February, 04 February, 28 March, 29 March, 29 August, 29 December 1881. 16 September, 16 October 1882. 03 September, 13 September 1883. 07 July, 02 August, 20 December, 22 December 1884. 11 February, 24 July 1885. 22 March, 04 May, 22 October, 27 December 1886. 15 February, 23 February, 01 July, 19 December 1887. 20 March 1888. 21 March 1901. 08 August 1902.

Dundee Advertiser: 07 April 1869 31 May, 12 June, 16 June 1871. 10 April, 11 September, 02 October 1879. 05 October, 08 October, 18 December 1880. 28 December 1881. 06 January 1882. 23 April 1884. 06 July 1885

Dundee Evening Telegraph: 28 August, September 30, October 11 1878. 02 July 1879. Tuesday 17 January, 27 November 1882. 19 March 1883. 15 February 1884. 10 February, 11 May 1885. 03 May 1886. 03 June, 25 August 1887. 26 July, 12 September 1888. November 08 1895

Dundee, Perth, Forfar, and Fife's People's Journal: 16 August 1873. 04 January, 08 November 1879. 10 July, 09 October 1880. 26 April, 30 August, 27 September 1884

Edinburgh Evening News: 18 August 1888

Evening News (Sydney, NSW: 1869 – 1931): 26 August 1872. 26 January 1882. 16 July 1884. 27 November 1889

Examiner (Launceston, Tas.: 1900 - 1954): 3 November 1904

Freeman's Journal: 18 March 1870, 20 December 1876. 17 April, 18 April, 15 December 1877. 22 June, 13 August 1878. 06 March, 23 December 1879. 07 October 1880. 21 October 1881. 07 January 1882. 18 November 1886. 23 June 1887. 20 July, 1888.

Fife Free Press, & Kirkcaldy Guardian: September 07, 1895

Geelong Advertiser: 24 January, 6 December 1884

Glasgow Herald: 19 June, 23 November, 29 December 1869. 18 March, 20 April, 23 August, 28 September 1870. 20 January, 12 June 1871. 01 January, 31 January, 13 February, 15 February, 19 April, 23 May, 30 July, 15 August 1872. 11 January, 05 April, 17 May, 28 August, 03 October 1873. 04 February, 22 May, 26 October 1874. 19 January, 23 September, 25 September 1875. 29 January 1876. 14 February, 28 February, 02 March, 16 April, 17 April, 24 October, 09 November 1877. 27 February, 28 May, 12 August 1878. 09 March, 21 April, 22 December 1880, 17 October 1882. 15 September, 17 September, 21 September, 09 November 1883. 06 May 1884. 22 March, 18 November 1886. 17 June, 18 May, 23 June 1887. 10 October 1890. November 22, 1895. 05 March 1897.

Greenock Advertiser: 24 February 1870

Hampshire Advertiser: 09 July 1887

Hartlepool Mail: 29 September 1885

Hull Daily Mail: 07 February 1894

Launceston Examiner: 28 January 1884

Leeds Mercury: 28 June 1869. 18 January 1871

319

Star of Greece

Liverpool Daily Post: 28 November 1862. 19 June, 21 June, 22 June, 22 July, 09 October, 01 November 1869. 23 February, 24 February, 19 March, 26 March, 20 April, 22 June, 23 August 1870. 20 January 17 October 1871
Liverpool Mercury: 06 February, 02 March, 20 April, 21 June, 28 June, 22 July 1869. 24 February 1870. 01 September, 03 October, 1873. 1877. 22 November 1876, 22 February, 07 March 1877. 28 August, 20 December 1878. 10 September, 02 October, 03 October 1879. 07 October 1880. 07 March, 03 May 1884. 20 February, 20 May, 11 June 1885. 24 December 1886. 18 May 1887
Liverpool Courier and Commercial Advertiser: 26 March, 22 June, 23 August, 24 August, 17 November 1870
London Daily News: 20 April, 23 July, 16 November 1870. 20 January, 22 May, 19 August, 21 August, 25 September 1871. 20 April, 06 December 1872. 11 January, 29 August 1873. 23 May 1874. 23 June 1876. 06 March, 03 July 1877. 12 September, 05 November 1879. 11 June 1885. 25 December 1886. 15 February, 17 December 1887
London Standard: 21 June22 June 1869. 19 March, 19 November 1870. 19 January, 20 January, 21 January, 20 November, 06 December, 15 December 1871. 20 April, 23 May 1872. 29 August, 14 October 1873. 21 September 1875. 28 February, 16 April, 07 November, 09 November 1877. 12 November 1878. 09 April, 12 September, 02 October, 05 November, 22 December 1879. 21 April, 07 October, 21 October, 04 November, 29 November 1880. 27 August, 19 October 1881. 15 September, 17 October 1882. 18 August, 17 September, 03 October, 06 October, 09 November 1883. 26 March 1884. 18 December 1884. 13 January, 14 January, 17 January, 10 February 1885. 27 February, 23 March, 25 March, 13 April, 15 April, 21 April, 03 May, 05 May, 18 November, 29 December 1886. 13 February, 14 February, 02 March, 19 March, 20 March, 12 June, 23 July 1888. July 21, 1894. August 20, 1895.
London Evening Standard: 21 June 1869. 24 February, 18 March, 18 November, 19 November 1870
Lloyd's Weekly Newspaper: 11 March 1877
Los Angeles Herald: 27 October, 28 October 1895. 23 September, 4 October 1908
Los Angeles Times: 18 July 1999
Manchester Courier and Lancashire General Advertiser: 23 October 1868, 08 March 1877, 10 September, 05 November, 22 December 1879. 29 August 1883. 26 January, 15 February, 25 September 1884. 09 July, 31 August 1885. 09 March 1886. 14 February, 18 May, 23 June 1887. 12 June 1888
Mataura Ensign: 9 February 1892
Morning Bulletin (Rockhampton, Qld.: 1878 - 1954): 30 May 1903

Star of Greece

Morning Post: 13 October 1881. 06 March, 15 December 1884. 02 April, 09 July, 31 August 1885. 27 February, 04 August, 25 December 1886. November 23, 1895. 14 January, 29 June 1887. 19 February 1900.

Newcastle Morning Herald and Miners' Advocate: 25 June, 7 July, 9 July, 11 July, 14 July, 17 July, 18 July 1885. 19 April 1888. Friday 7 July 1905. 4 May 1906.

New York Times: June 03 1878. November 4, November, 26 November 30, December 21, December 27, 1879.

New York Herald: June 2 1878. November 4, November 30, December 20, 1879. December 17 1884.

New York Tribune: March 17 1877.

New York Evening Herald: December 31 1888.

News (Adelaide, SA: 1923 - 1954): 17 July 1925, 17 November 1927

New Zealand Herald: 21 March 1887

Northern Whig: 21 September 1868. 20 April 1870

Nottinghamshire Guardian: 31 August 1883

Observer (Adelaide, SA: 1905 - 1931): 7 December 1929

Otago Daily Times: 22 July 1878. 18 February, 8 May, 9 May, 2 June 1885. 8 February, 9 February, 16 February 1892.

Otago Witness: 19 February 1876. 27 July 1905.

Pall Mall Gazette: 22 October 1878

Poverty Bay Herald: 17 July 1905

Public Ledger and Daily Advertiser: 16 March, 07 April, 13 April, 21 June, 22 June, 14 December 1869.

Queenslander: 26 June 1875

Reynolds's Newspaper: 11 March 1877

Queensland Times, Ipswich Herald and General Advertiser: 29 June, 13 July 1875

Sacramento Daily Union: 28 September 1895

San Francisco Call: 6 January, 27 May, 30 May, 22 July, 27 September 1894. 7 July, 7 September, 12 September, 29 September, October 02 1895. 29 July 1897. 1 February, 9 February, 15 October 1898. 4 February, 2 January 1899. 23 September, 24 September, 3 October, 4 October, 7 October, 14 October, 17 October, 23 October, 27 October 1908

Sausalito News: 14 September 1895

Shields Daily Gazette: 18 January 1871. 19 April, 20 April, 28 December 1872. 20 January 1875, 07 December 1878. 05 March, 06 May 1879, 24 December 1886. 21 July, 23 July 1888. September 26, 1894. August 21, September 07, 1895. 30 January 1896. 19 February 1900. 21 March 1901.

Sheffield Daily Telegraph: 23 June 1887

Sheffield Independent: 18 November 1878

South Australian Advertiser: 10 December, 11 December, 17 December, 24 December, 28 December 1880. Thursday 28 April 1887. 26 March, 18

Star of Greece

May, 12 June, 18 June 7 April, 14 July, 16 July, 17 July, 18 July 1888. 15 March, 1 November 1889. 28 August 1901. 30 May, 5 June, 25 June 1903. 7 June 1905. 10 May 1932. 22 August, 3 September 1934, 13 July, 14 July, 20 July 1938. 4 April 1940. 3 July 1941. 9 November 1944. 10 September 1945. 29 November 1946. 28 April, 29 April 1952. 21 April 1954.

South Australian Register: 29 November, 21 December, 28 December 1880. 12 February, 24 March, 25 March, 30 March, 7 April, 12 May, 13 May, 14 May, 18 May 1881. 8 July 1885. 9 March 1886. 11 May, 12 June, 26 June, 27 June, 28 June, 2 July, 14 July, 16 July, 17 July, 18 July, 19 July, 20 July, 21 July, 24 July 1888. 14 August, 19 August 1889. 28 January 1891. 22 April 1897.

South Australian Weekly Chronicle: 12 July, 6 September 1884, 30 June, 14, 21 July 1888.

Southland Times: 17 February 1892

Star (New Zealand): 7 January, 17 June 1885. 15 February 1892. 2 November 1895.

Sunderland Daily Echo and Shipping Gazette: 22 May 1874. 25 September 1875. 24 June, 26 June, 05 September, 19 December 1876. 09 January, 02 March, 06 March, 16 April, 17 April, 09 November 1877. 26 February, 29 October, 19 December 1878. 11 September, 01 October, 02 October, 03 October, 15 December 1879. 07 September, 06 October, 07 October, 22 December 1880. 30 August 1881. 25 November 1882. 26 September 1884.

The Argus: 21 March, 30 August, 13 October 1881. 18 January, 28 February 1882. 24 January, 25 January, 26 January, 24 March, 26 March, 27 March, 28 March, 5 December, 6 December 1884. 14 February 1885. 24 February 1899. 29 June 1903.

The Inquirer and Commercial News: 26 September 1890.

The Mail (Adelaide, SA: 1912 - 1954): 22 March 1924. 27 June 1953.

The Mercury (Hobart, Tas.: 1860 - 1954): 24 July, 26 July 1905

The Register (Adelaide, SA: 1901 - 1929): 30 May, 29 June 1903

The Sydney Morning Herald: 24 March 1881. 17 January, 25 January, 25 February, 27 February 1882. 7 July, 9 July, 18 July 1885. 18 April, 12 June 1888. 23 June, 29 June 1903. 11 July 1905. 11 September 1923. 20 May, 22 May 1929.

The Sydney Mail and New South Wales Advertiser: 4 March 1882. 11 July 1885

The Telegraph (Brisbane, Qld.: 1872 - 1947): 17 January 1882

The Times (London, England): October 23, 1868. July 05, 1869. February 18, March 04, March 18, Mar 19, November 29, 1870. September 04 1871. Dec 06, 1872. April 23, 1873. October 11, 1875. February 28, April 17, November 09, 1877. February 27, November 12 1878. December 22, 1879. April 21, October 07, October 15, October 18, Dec 16 1880. January 29,

Star of Greece

August 27, October 13, October 14, October 19 1881. August 18, Sep 15, October 06, November 09 1883. January 13, January 14, January 17, May 11 1885. Aug 04, 1886. May 18, December 17 1887. June 14, July 23 1888
Wallaroo Times: 18 July, 21 July, 25 July 1888
Waikato Times: 9 February 1892
Wanganui Chronicle: 16 May 1877
West Australian: 26 September 1890. 24 June 1897
West Briton and Cornwall Advertiser: 03 September, 14 September, 27 September, 28 September, 24 October, 06 November, 08 November, 09 November, 13 November, 14 November, 21 November, 23 November, 26 November 30 November, 03 December, 05 December, 12 December, 14 December, 18 December, 20 December, 29 December 1877
Western Daily Press: 29 August 1883. 07 March, 13 March 07 July, 14 July 1884
Western Mail: 22 June 1869. 21 April 1870. 10 September 1879. 28 March 1885. 2 August 1890. November 22 1895. 9 January 1930.
York Herald: 17 October 1881. 02 November 1878, 13 February 1888. 27 June 1894.
Yorkshire Post and Leeds Intelligencer: 30 June 1887. 22 December 1879. 02 May 1884. 18 November 1886. 17 May 1887. 11 July 1905.

Websites

Elephind.com: Search the world's historic newspaper archives. https://elephind.com/
Chronicling America « Library of Congress. https://chroniclingamerica.loc.gov/
Books Boxes & Boats Maritime & Historical Research. https://www.maritimearchives.co.uk/
FamilySearch. https://familysearch.org/
Maritime history archive. (n.d.). Memorial University | Newfoundland and Labrador's University | Memorial University of Newfoundland. https://www.mun.ca/mha/
California digital newspaper collection. (n.d.). https://cdnc.ucr.edu/cgi-bin/cdnc
National Library of New Zealand. (n.d.). Papers Past. https://paperspast.natlib.govt.nz/
Crew List Index Project. https://www.crewlist.org.uk/
Mariners and ships in Australian waters. (n.d.). Mariners and Ships in Australian Waters. https://marinersandships.com.au/search.htm
University of Oregon, Knight Library. (n.d.). Historic Oregon Newspapers. https://oregonnews.uoregon.edu/
Ancestry® | Genealogy, Family Trees & Family History Records. https://www.ancestry.co.uk/
History Trust of South Australia. https://history.sa.gov.au/
Trace your Family Tree Online | Genealogy & Ancestry from Findmypast. https://www.findmypast.com/
Trove. https://trove.nla.gov.au/

Star of Greece

About the author.

Paul W. Simpson is a writer, historian and educator. He is the author of several books on maritime history including Star of Greece – For Profit and Glory, Windjammer – Tales of the Clipper Ship Loch Sloy, Neptune's Car – An American Legend, and The Last Captain. Paul grew up on Kangaroo Island, the last place where square-rigged ketches sailed commercially and where tales of shipwrecks were part of the local folklore. He has spent the last decade researching and writing books about 19th century clipper ships and the folk who sailed in them. When not found trawling through musty old books and newspaper archives, he can be found combing windswept beaches for treasures with his daughter and fellow adventurer.

Index

Star of Greece

Star of Greece

Star of Greece

L - #0181 - 040920 - C0 - 210/148/20 - PB - DID2900701